Producer's Guide To Interactive Videodiscs

Martin Perlmutter

Knowledge Industry Publications, Inc.
White Plains, NY

The Video Bookshelf

PRODUCER'S GUIDE TO INTERACTIVE VIDEODISCS

Perlmutter, Martin

Library of Congress Cataloging-in-Publication Data

Perlmutter, Martin.
    Producer's guide to interactive videodiscs/Martin Perlmutter.
        p. cm. -- (The Video bookshelf)
    Includes bibliographical references and index.
    ISBN 0-08-679173-7 : $45.00
    1. Interactive video. I. Title.   II. Series.
    TK6687.P47   1990   90-19233
    004.69--dc20   CIP

ISBN 0-86729-173-7

Copyright © 1991, Knowledge Industry Publications, Inc.,
701 Westchester Avenue, White Plains, NY 10604

Printed and bound in Canada

10   9   8   7   6   5   4   3   2   1

Table of Contents

List of Tables and Figures

Foreword

God, they say, is in the details. So is Marty Perlmutter. In all my years as a journalist covering interactive video and related topics, I have never seen such a comprehensive look at the subject—a subject that is clearly the most exciting and significant technology/technique ever to spring from that miracle-monster, television. Video has become ingrained in our day-to-day lives—giving us directions on touch screens, printed training manuals, instruction books and sales brochures. Likewise, interactive video is increasingly as common and as integral to everyday life as modem communication, fax and those coffee makers with automatic timers that have your coffee waiting for you when you wake up in the morning.

Marty Perlmutter has a unique handle on all this, as an interactive video producer and pioneer. His *Producer's Guide to Interactive Videodiscs* shares—generously—over a decade's worth of experience in the field, with no stinting on details or tricks of the trade. On top of that, he has written in a friendly, highly readable, conversational style—wonderful proof that the most complex high-tech subjects in the world can be clearly explained. This is an amazing book.

That, however, is to be expected from Marty. When I first met him, I was handing him an award—*Video Review* magazine's VIRA statuette for Best Interactive Program, given to him for producing the ground-breaking interactive MysteryDisc, *Murder, Anyone?* As the magazine's resident "expert" on this esoteric (c. 1983) subject, I mounted the stage and quipped to the tuxedoed dignitaries, "Interactive video? What's interactive video?" Marty took that as his cue to give a brief eloquent explanation that made all of us in the crowd feel as if we were standing on the cutting edge of a fabulous, science-fictiony new medium. Which, of course, we were and still are, and will continue to be. Interactive video is at a marvelously fertile and organic state: a toddler absorbing new bits of knowledge at an impossibly rapid pace. As anyone with a child knows, there is no more magical and exciting a time than when any future at all that we can imagine is still a very real possibility.

In terms of interactive video's future (and present and past), I can't envision a more practical, step-by-step guide than this. It has everything, from the proper way to handle still frame data to the built-in moral implications of this manipulative new medium. And it is enlightening, in the way that the best discourses on a new technology can be.

Bravo!

Frank Lovece

Acknowledgements

This book has a family. It has ancestors and parents. It had a midwife. And, someday, I hope it will have children.

Its grandfather was Marshall McLuhan, whose probes into the perceptual and cultural mechanics of media awakened me to the importance of user-controlled video. Its godfather was Gene Fairly, mentor and friend, who introduced me to interactive videodiscs and allowed me to practice the craft of creating them. His colleagues of videodisc publishing, Ann-Marie Garti and Margaret Bates, were generous and supportive siblings to this child of my spirit.

Ed Rosenfeld, 21st Century encyclopedist, introduced me to Gene Fairly, and so to videodisc. David McCall kept me from stepping on most of the land mines that littered my path in learning the IVD producer's art. They are the book's, and my, brothers.

Mia Amato was its midwife. She caught it, in its earliest incarnation, and helped to give it form and substance. Linda Richmond, beautiful sister, typed the first draft onto diskette.

This book's mother is Dr. Bob Huff. He nurtured it, suckled it, and tenderly raised it, from rickety toddler to robust youth. He believed in it when I did not, and edited it and printed it out, again and again, until it came to believe in itself.

This book had teachers—Dick Cavagnol, Brian Kahin, Eric Roffman and Peter Trachtenberg. It has a gang in its adolescence—technologically street-wise and witty guides in Richard Haukom, Garry Hare, Mark Heyer and Tom Hargadon—my Green Street Gang. And it had an artist friend in George Nachtigall, whose laughing drawings adorn page after page.

It has sweethearts in my wife, Miki, and my daughter, Sasi, who tolerated it and cherished it as it grew.

And now, this child all grown seeks a wider circle. You, dear reader, are part of its family.

Introduction

So, you found a client who's heard about interactive TV. And now they think they've got a line on video's Holy Grail: the perfect answer to every training, marketing, customer support and point-of-purchase need. Now they want you to make a videodisc to solve their every problem. And you told them you could do it. You said, "Sure, I've got disc production experience! Lots of it." And you've found out that you're going to have to wing it because there is no literature in the field. Well, relax. In the next few thousand words, you will find much of what you need to know to bid, design, shoot and premaster your promised interactive television product.

Normal video or film productions just begin and end. Maybe a user will pause in the middle, to take a quiz or get a cookie. Videodiscs are different. In viewing a disc, the picture may abruptly and auto-

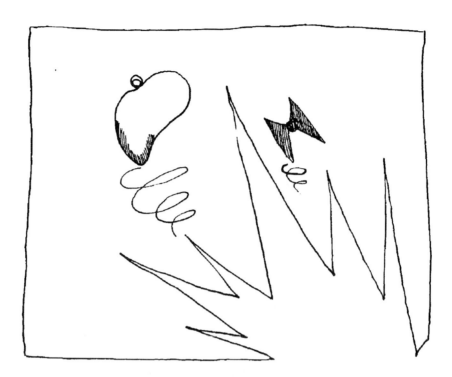

matically halt at several points. There may be thousands of still-picture frames, each accessible by individual number, through which the viewer can browse or hunt. Computer-generated text or graphics may appear on the screen, either during motion sections or over still images. Separate sound tracks may be used to give different levels of information to disparate audiences, or different clues in a mystery story. There may be 16 unique ways to see the footage found on the disc, or there may be a million. Discs "branch," as the computer folk have termed it, depending upon choices the user makes. Videodiscs can also freeze on a single frame or a sequence of stills, then tear off within seconds to the farthest reach of the disc to launch into another motion sequence or still.

"So what?" you ask; videodisc productions are supposed to be expensive. Can they justify their costs? How do you sell such a project to a client? And, how do you make good on the promise to deliver one?

Interactive video is destined to play a major role in both training and entertainment. Indeed, the futures of video and of computing will be closely intertwined. Most of what is unique about producing for videodisc remains true whether one is creating imagery for delivery on interactive tape, CD-ROM or some as-yet-unknown means of retrieving motion pictures from solid state circuits.

The videodisc beckons because it is the high country of interactive television. It offers, for the present, the best in access speed, durability and density of pictorial storage for retrieval of both still pictures and motion imagery. One need not love the disc or believe in its economic viability to seize upon it as a training ground for learning the producer's skills for creating programs that interact with the user.

Interactivity means that the product is under the user's control. This calls for a new, inside-out sort of program design. We must climb inside our no-longer-passive viewer's mind and think through all of the possible directions the program may go in order to respond to the user's needs or commands.

The rules we adopt in mastering this process should transfer neatly from disc to CD-ROM to chip. However we call up the pictures, someone has to design them, sequence them and create the control program that makes the user feel as if he or she is running the show.

This book is about those rules and related tricks. Our focus will be primarily on the videodisc, because that is the delivery system today. Soon the successors to disc will appear. Producers of interactive videodiscs will be prepared for that future. They will grasp the scheduling, time- and budget-management exigencies and production peculiarities of the interactive delivery environment. And, they will be positioned to shape a future wherein formerly passive viewers become the controllers, perhaps even the creators, of their video experience.

"To know one's ignorance is the best part of knowledge." — Lao-Tzu

I like to joke that *producing is never having to say you're sorry.* If it's true that mistakes are costly in normal "linear" teleproduction, it is ten times more true of interactive productions. Videodiscs that are poorly conceived or inadequately pre-produced tend to die silently, and expensively.

This book will attempt to set forth the unique characteristics of interactive videodisc production. It will explain how to plan and execute successful videodisc projects. We will start with an outline of the basic features of the videodisc and the levels of interactivity available to users, then examine some proven procedures for designing and managing interactive videodisc productions. Along the way, we will also cast a glance at some future directions for exploration and development.

The goal in all of this is to underline the differences between linear or non-interactive productions (ones that simply start and stop) and interactive ones. In so doing, we hope to equip producers and media managers with a checklist of skills and support services that should be at hand before embarking on a production venture in this new realm.

Armed with knowledge of the distinctions between interactive and linear uses of video, and duly warned about the import of these, I am hopeful that you need never be sorry, or even mildly panicked, on your path through these video highlands.

In the near future, there will be a panoply of potent tools to assist us in our visualization tasks. Storyboard programs on our lap computers, and dynamic script programs that tie into our project management flowcharts will make our videodisc production projects come alive—before the final product is to be delivered.

For now, we must frequently avail ourselves of humbler resources. In the following chapters, you will encounter some methods that work today. But whatever the state of our job aides, producers of interactive software must invest their energy throughout industry and education to help achieve the glistening promise of the videodisc. If we do not stretch, and survive this hungry epoch, we may have to await high-bandwidth, fast and affordable computer chips to do the work that we might do today, had we but the vision, and the guts.

Marty Perlmutter

Part 1
Theory in Practice

1 Adventures in Discland

THE EARLY YEARS

Back in 1979, when the disc was a gleam in MCA's eye and a business plan on some IBM officer's desk, there truly were no software providers. MCA's DiscoVision subsidiary came up with a plan to seed the new technology with some worthwhile software of an interactive variety to spice up the rather dull menu of Universal Studios' film offerings. The result was called Optical Programming Associates (OPA), a group of not-terribly-well-funded individuals with contacts and some ideas for product in the new delivery medium. One of OPA's first projects was the creation of *The First National KidDisc*—still an excellent item for acquainting newcomers to disc with the medium's potential. The *KidDisc* first demonstrates, then uses, virtually all of the features available on the Level 1 consumer player: dual sound tracks, fast motion, slow motion and frame advance, and it integrates these capabilities into a delightful series of activities for children to enjoy. A user can test himself or herself on the flags of the world, learn origami, knot tying or folk dancing, take a flight over Los Angeles, or play an elegant little game that challenges one to stop on a single bull's-eye frame while the disc is advancing. The *KidDisc* is a simple but elegant *tour de force*, and although OPA has long since ceased operations, the *KidDisc* is still available in stores.

The CIA and the military, along with certain auto companies, were early customers of interactive video. Companies such as GM and Ford, purchased thousands of disc players and established networks for disc-delivered dealer training and customer education. It could be said that the auto industry's film-based discs kept the interactive videodisc (IVD) alive during its early days in the United States.

In the early 1980s, more serious applications of disc came under consideration. One

3

project, directed by David Hon for the National Heart Association, led to the creation of an interactive training disc coupled with a life-size dummy to train people in cardiopulmonary resuscitation. The now-famous *CPR Disc* system is a high-water mark in the development of packages that synthesize discs with computers and simulation environments. A computer monitors dozens of sensors within the dummy, which comes in both adult and child sizes. Too much pressure or not enough and the computer tells you that you've made a mistake, tells you what you did wrong, and then plays a short disc sequence that shows you how to do it right.

By 1982, a growing cluster of companies and individuals had seen the possibilities of disc and were on the move. In October of 1981, Pioneer LaserDisc had been spawned by Pioneer Electronics in Japan and a major new player was added to the field. IBM was discovering, rather painfully, that real profits in disc were perhaps a decade away, and DiscoVision Associates' joint venture with MCA folded. But new faces began to crowd the scene. In Cincinnati, a small group of financiers calling themselves VidMax initiated two important projects: (1) a disc that contained the entire painting collection (plus a tour) of the National Gallery of Art in Washington, DC; and (2) the MysteryDisc.

The first MysteryDisc, *Murder, Anyone?* was the brainchild of several talented sires. (See Figure 1.1.) Gene Fairly of Videodisc Publishing directed VidMax to Norman MacFarland, head of the design department at the University of Illinois. MacFarland, an accomplished game designer, structured the disc to have 16 different stories and a like number of correct solutions. He made impressive use of all the disc's capabilities, including chapter stops, searches to individual frames and separate audio tracks. The result of his efforts was a disc with 16 unique one-hour games stored in 30 minutes of linear space. Hy Conrad, a theater and mystery writer, took Norman's design and fleshed out a wonderful weave of characters and plots. From his efforts emerged a script that was as thick as a Manhattan phone book. On a day in the spring of 1982, that daunting script was presented to a bewildered group of film and video producers who sought to bid on the project. Four months later, after some epic efforts, Ghost Dance Productions delivered the edited pre-master of the MysteryDisc. It remains the most popular interactive product on the consumer market. More important, however, it suggests the tremendous range of possibilities for non-computer-driven applications of videodiscs.

Medical applications of disc began to flourish at about the same time. Certain pharmaceutical companies saw the disc as a potential tool for training as well as a potent vehicle for marketing—if only they could get such programs delivered to their harried audience of medical practitioners. Dr. Campbell Moses, of Medicus Intercon (a company that produces and promotes medical products), conceived and implemented a system of interactive training centers—the Miles Learning Centers—which were installed in 280 teaching hospitals across the country. In these Centers, new discoveries and medical techniques could be presented in a dedicated location at the hospital site. Trainees could also be tested at the user's convenience.

The Miles experiment, which was still in place in the late 1980s, foundered on a

Figure 1.1: The First MysteryDisc, *Murder Anyone?*

Photo courtesy of VidMax.

shortage of worthwhile programming. The creation of a training network takes more than installation of a base of machines and supply of a few pilot programs. To effectively create such a pipeline, a massive effort is necessary, one that probably requires the joint efforts of several institutions and that goes far beyond the scope of a basically promotional effort by a single drug company.

Games were the watchword by the dawn of 1983. A surprising newcomer to the then-booming arcade market, *Dragon's Lair*, captured the imaginations and the quarters of millions of ardent videophiles. At the console of this first videodisc arcade game, the player used a joystick to tell an animated figure on the screen, Dirk the Daring, to go forward, move to the side, go back, or swing his sword. The instruction took up to three seconds to result in a visible response on the screen—usually the untimely demise of the hero. The disc had to scoot off and find the appropriate next sequence that resulted from the player's decision—such as Dirk falling into a chasm. The screen blanked out for this interval. Only the truly faithful found this a satisfying process, at fifty cents a game. Arcade owners embraced the surge in revenues, however momentary. Ten thousand of the games were sold, and the conventional wisdom, for a season, was that interactive videodisc arcade games were the way to go. At the AMOA (Automatic Machine Operators Association) arcade games convention in October of 1983, no fewer than 14 interactive videodisc games were shown, but only half a dozen actually went into production. The game consoles cost roughly $4,500 apiece—about twice what the most expensive traditional arcade games cost. Then, suddenly, the entire arcade industry went into a nosedive. Bally,

Williams and other companies lost millions of dollars on their disc investments.

At about the same time that the interactive game market was sinking, RCA folded its tents on its always ill-fated venture into disc. In the early 1970s, RCA had developed a contact stylus disc player—the capacitance, or CED, disc machine. By 1980, when this movie playback device was released to the consumer market in one of the biggest promotional blitzes ever, it was not well received. Only a few consumers bought it, and since RCA made only the most feeble pass at offering or promoting interactive software products for its player, the massively financed promotion of this machine utterly obliterated public consciousness of disc's unique capacity to offer the user interactive control of the program content. Without doubt, RCA's ill-advised entry into the marketplace, which lost them $600 million by the time they quit, was a major contributor to disc's lack of success in the American consumer market.

Meanwhile, in Japan, Japan Victor Corporation (JVC)—holders of the patent on VHS videotape recording—and Pioneer were having some success finding a consumer market for their different-standard disc players. Two factors helped to drive sales of disc machines forward in Japan: first, the Japanese really cared about the visual quality of their programs, so they gravitated toward the higher quality playback offered by disc; second, there was a sudden and manic surge of popularity in the indoor sport known as "karioke," also called "music-minus-one," which permitted untrained people to stand at microphones in bars on the Ginza and other popular social gathering places and sing popular songs along with the recorded background music. It was not uncommon for a bar entering the Karioke market to order up to 100 discs to go along with its new VHD (very high density) or Pioneer laser disc player. This mania has receded a bit, but is still a very popular entertainment in Japan, and constitutes an important part of that disc market.

Back in the United States, the post-arcade doldrums settled around the disc business in early 1984, and training became the one vital segment of the industry. The consumer market was dead. Unit sales of players were about 6,000 a month, but big companies, such as IBM and AT&T, were beginning to make heavy use of IVD for internal training and for customer support and sales. Digital Equipment Corp. (DEC) made extensive use of disc in its IVIS (interactive video instructional system) training consoles, used internally for technical training and offered to the outside world at a staggering five-figure price per IVD workstation. For a while, there was a healthy market for design and production services for these big firms—and then even that dried up. As far as I know, there was no 1985 in the disc business in the United States.

POP, TRAINING, LEVEL 3 AND INSTANT BRANCHING

The year 1986 dawned with some promise. Pioneer LaserDisc (now LaserDisc Corp., a division of Pioneer Electronics) developed a stunning catalog of interactive products that were available to its Japanese customers. Offerings included CPE (computer program encoded) discs of all the old American arcade games (now playable in homes with MSX computers) and collections of the special effects of some of the great wizards, such as Robert Abel. A growing number of corporations voiced interest in IVD for training. The federal government talked up Level 3 for military training—although actual government spending on the technology grew non-explosively. Point-of-purchase (POP) systems, which were supposed to be big by 1985, but never were, began to bubble and boil. Banks began to use discs to train their customers in ancillary services, and the Japanese market continued to grow: one million units at year start; nearly two million by year end.

Also that same year, a medical training disc, *The Case of Frank Hall*, was produced for the National Library of Medicine. The disc featured voice synthesis of phrases such as, "Where does it hurt?" which were followed by a complex branching network of diagnostic possibilities and, finally, the correct diagnosis of the case. Other discs created in 1986 include a directory of jobs and professions for the U.S. Department of Labor produced by Fusion Media; and high-impact exhibits such as the Boston Museum of Science's *Water Crisis* and *Nova Science Quiz*, which helped to showcase the possibilities of interactive videodiscs in museum environments.

In 1987, instant branching arrived in arcade gameware with the advent of *Freedom Fighter*. This game had all the technical panache lacking in *Dragon's Lair*. It offered double-wide pixels, which permit a user to pan to the left or right of the screen edge for additional views from the same recorded frames of video. Instant branching meant truly instant changes of the image, depending on user inputs, through the magic of *interleaving*—that is, laying down on disc successive frames of different movie sequences and letting the user select which to watch. Access times of advanced disc players had become instantaneous over short jumps, so the capability for truly instant response was at hand.

DIGITAL VIDEO INTERACTIVE (DVI)

Perhaps the biggest development in interactive video in 1987 was the unveiling of DVI (digital video interactive) at that year's CD-ROM Conference in Seattle, WA. A cadre of scientists and software engineers that had worked on RCA's brief foray into interactive CED disc production had been quietly and steadily at work on a major breakthrough in video compression/decompression technology. At the CD-ROM gathering, these survivors of RCA's David Sarnoff Research Center, recently acquired by General Electric (and then given, but for this one project, to SRI International), demonstrated the capability to read full-motion video off a CD-ROM data disc. This had been an unthinkable miracle up to that moment. The crowd rose to give the feat a standing ovation, and a vast new capability for generating computer video had been born. With 72 minutes of motion video, windowing capabilities, digital effects built into the software, and continued improvements in the video decompression algorithms promised, DVI seemed to be the Holy Grail of interactive computer visual technology.

It was also in 1987 that the Criterion Collection of enhanced Level 1 videodiscs began to receive commercial acceptance. *King Kong* and *Citizen Kane* were released by Bob Stein and associates with commentary by directors and special-effects craftsmen, theatrical trailers, storyboard sketches and similar enhancements for the afficianado. Stein's vision of the videodisc as "a kind of book" took on vivid multimedia meaning in this product; here one finds a virtual video textbook. And, in that same year, the National Geographic Society began work on Geographica, an interactive science center for geography in the Explorers Hall at National Geographic headquarters in Washington, D.C. This project includes Earth Station One, a 72-seat interactive amphitheater featuring an 11-foot globe that moves, projects images and responds to inputs from keypads on viewers' seats.

COMPACT DISC-INTERACTIVE

In 1988, the Interactive Video Industries Association came into being, with plans to create an interactive applications gallery—Tech 2000—to be built adjacent to the Washington, D.C. convention center. Currently open, Tech 2000 offers hands-on encounters with nearly 80 interactive products. It is a powerful and unique exhibition.

Philips Corp., in concert with its wholly owned record company subsidiary, PolyGram, had incorporated American Interactive Media to fund the development of titles for CD-I (compact disc-interactive), which was then scheduled for consumer release in 1990. Growing out of the wild consumer enthusiasm for compact disc audio, Philips (the patent holder) sought a new profit stream from optical multimedia. CD-I was born of this rosy glow. The key, Philips rightly reckoned, was to fund software development for the proposed product so that consumers might have some reason to buy it—a lesson still unlearned in videodisc.

In 1987 to 1988, one could see the first products created for DVI. Perhaps most impressive were Bank Street College's *Palenque* demo and Videodisc Publishing's *Design and Decorate*. *Palenque* allows a user to take a joystick-guided walk around a Mexican archaeological site, calling up experts, taking snapshots, retrieving maps and flying up into the jungle canopy. Here is a *tour de force* of the new medium's multimedia capabilities. *Design and Decorate*, allows the user to define a room and then cover its walls in any color or pattern, install furniture, cover the furniture, add rugs and paintings, and then take a walk about. This is all achieved digitally, and is a stunning accomplishment both in digital imaging and in point-of-purchase experience enhancement.

The prolific Videodisc Publishing, Inc. (VPI) released *Andrew Wyeth's Helga Pictures* in 1988, featuring a tour of the Wyeth farm, an examination of four techniques the artist used and an analysis of the place of this work in contemporary art. Also that year, LaserDisc Corporation released a two-volume *Anthology of American VideoArt* that attracted a Culture/Hobbies/Science Award at Tokyo's Audio-Visual Age show. Training and point-of-purchase productions continued quietly. The American disc consumer market remained anemic, though now videodiscs could be found in large video stores. Intel acquired the DVI technology, meaning that it would soon find its way into new-generation IBM PCs.

Figure 1.2: Mark Heyer of Heyer-Tech with Mac-Driven Panasonic OMDR

Photo courtesy of Heyer-Tech.

OPTICAL MEMORY DISC RECORDERS AND THE FUTURE

In 1989, IBM, Microsoft and Intel joined forces behind DVI. The first, demonstration phase of the technology was complete, and the real work on the processing chips and authoring tools began. With the cost of a development system at $20,000, DVI headed for the great techno-closet in the sky for a few years. Meanwhile, CD-I demonstrated full-motion and full-screen capabilities, and more and more titles made their way through the tortuous American Interactive Media pipeline—there were 30 or so in development by mid-year. Pioneer began to sell a $700 combo videodisc player—capable of playing compact discs or 12-inch laser discs, and videodisc player sales in the United States began to swell. But just as it looked as if the technology was about to take off, along comes a whole new ballgame. Now you can roll your own discs! First Panasonic, then Teac and now Sony offer optical memory disc recorders (OMDRs) with WORM (write once read many times) drives (see Figure 1.2). For a scant $16,000, you can buy a desktop device that will record up to 54,000 frames of video on one side of a $260 blank disc. The OMDR allows you (or your client) to make a recording of available training software for which you write a control program and presto, you have IVD. Players for these home-brewed discs start at about $3,500—so we're talking about a new way to produce and distribute interactive videodiscs. Meanwhile, at NeXT Computer Corp., a Canon erasable disc lies at the heart of the beast. We have only begun to hear of this make-your-own-disc phenomenon. Someday, there may be no other kind.

By 1990, Pioneer and KDD (the long-distance phone company of Japan) were at work on an inexpensive erasable optical drive. Intel was shipping under $2,000 DVI boards to run video on PC screens, and two major new compression technologies were announced. CD-I promised full-motion video in its base player, and heaven be praised, Pioneer's $495 combo CD-laser video players were selling like hotcakes. Fasten your seat belts.

2 Getting the Most From Videodisc's Unique Characteristics

In most cases, videodiscs are not intended for normal linear playback. By definition, an interactive videodisc (IVD) invites its users to become participants, selecting still pictures or motion sequences according to their desires or the program's instructions. The creation and placement on disc of materials that may not be played back in the same order in which they were recorded is a unique enterprise. In creating interactive videodiscs, there are techniques of design, scripting, shooting and editing that depart from standard production practice.

The creation of an interactive disc, even in a mechanical sense, entails much more than the preparation of stills and short motion sequences for an unpredictable playback arrangement. That approach might work if your disc simply has an archive on one side and a linear story on the other. But it may well fail in a training or sophisticated game application.

Creating a magical map through unexpected experiences, as one might in a game, cannot result from mere recombination of formerly linear elements that are strung together. To make interactive discs work, you must feel the play and interplay of all the pieces that make up the experience. It is vital for the disc-crafter to foresee the special quality of the interactive user's experience, or the results will deviate from expectations.

From the producer's perspective, the difficulty of previsualizing a disc product is compounded by the reality that there are few tools for assisting in that task. Left to the resources of a flowchart or even of a *HyperCard* stack (an interactive authoring system from Apple Computer), it takes a special imagination to see the play of a game or the

11

impact of a training sequence and its remedial branches. Even after these elements are, hopefully, properly produced and arranged on tape and ready to be mastered into the interactive product, there is still no way to really see the product in action. One must wait until it is too late to see if it works—until the money has been spent, the schedule lived through, and the final delivery system fully configured and operating. Then, and only then, does the interactive producer know whether his or her efforts have borne fruit. I do not overstate the case here. This is literally true, and it represents a real challenge to those who would author for this emergent medium.

The tendency, in the face of this daunting invisibility of product, is to shoot for low marks. In disc pioneer David Hon's words, "Producers either do a lot of showy linear material, or if they're into computer programming, there's always death in the still frame." Why stretch for some uncertain and difficult-to-visualize result when one will not know if one has succeeded until it is too late? Wouldn't it make more sense to simply settle for, say, a visual database and then just massage it with an outboard computer—and not worry about creating some high-impact simulation of a product's operation or a role's actual feel, out there in the world of blood and guts? The answer is: Be brave.

Again, in Hon's words, "The failure of most videodisc projects can be accounted for in two words: No guts." The trick is that you must risk much in order to achieve an interactive relationship—a relationship in which producers and programmers relinquish their power: the one-way power of the medium over the user. "This new interactive relationship delivers power to the people. That's what *Space Invaders* brought us: Real power to the people!" says Hon.

To deliver the goods, we're going to have to stretch. Ultimately, we will attempt to create "ah-ha" experiences for the users of our products. We will guide them to moments of sudden and memorable clarity, when a process or idea suddenly becomes perfectly transparent and forever after usable in another context. With bold and intelligent use of interactive programming, we can allow our program users to arrive at unique syntheses of our inputs and so achieve what they perceive to be personal discoveries. There is no better way to help people to "own" newly acquired knowledge.

APPLICATIONS AND THE VISUAL EXPERIENCE

Within a decade, most FORTUNE 1000 companies will be doing training and customer service using videodisc (or some CD-based computer visual system). IVD, along with its computer-visual brethren, is the medium for cost-effective, self-paced and individualized delivery of information. The videodisc, when combined with a computer, offers the ultimate in data-rich and user-responsive teaching or simulation systems. It is ideal for training applications that involve the use of case studies or role modeling. Videodisc far outstrips the capabilities of computer-based training (CBT) for any application requiring human interaction skills or illustration of dynamic processes.

Videodisc offers flexibility in addressing what may be a segmented audience. Its vast information storage capacity, combined with rapid access speed, multiple audio tracks and integration of still pictures with motion sequences, offers a vast array of tricks for cramming information into available space.

Videodisc has been shown to be at least twice as fast as teacher-lead instruction in bringing management and sales trainees through their preparatory processes. Disc releases training from the constraints of a particular trainer's skills or system knowledge. It also allows the trainee to work at his or her convenience, and can facilitate growth by bringing new hires up the learning curve at a faster pace. By reducing instructor and trainee time costs, and by evening out the instructional level, IVD lowers the costs of training.

Video is also a familiar medium, which makes it ideal for providing positive role models and processes, and for maximizing user involvement in problem solving. The utilization of graphics can further add to the impact and memorability of a product, in ways usually unachievable when done in person or via computer-based training.

By integrating the friendliness, warmth and role-modeling characteristics of video with the tracking and simulation capabilities of the computer, you can achieve the most advanced and comprehensive training vehicle available. It features flexibility, user-focus, verisimilitude, security, cost-effectiveness, impact enhancement, updatability, and multilingual capability—benefits, numerous and real.

When working with a client, be frank about the constraints as well as the benefits of IVD. It isn't computer-based training and shouldn't be asked to do what CBT does already, and does well. If you can get by with text screens, don't use videodiscs. If sequenced stills will suffice for your sales or training purpose, consider CD-ROM as a storage and retrieval medium, or just use micrographics out of a PC. It is unwise, as well as unethical, to lure a customer to disc when cheaper or easier solutions will suffice. To quote an apposite Chinese proverb: "One fish can muddy a whole pond." Disc has already been muddied by ill-advised ventures and inappropriate applications. Remember, IVD is video. Use the full-motion capabilities, as well as the still-store resources it offers. Simulations, role-modeling and real processes played out dynamically and interactively—these are the showcases for interactive video.

It is as challenging to previsualize a juicy, creative and appropriate use of IVD as it is to create one. In fact, to date, there are precious few applications of videodisc that have made outstanding use of the technology. The two biggest users—the U.S. military and the automotive industry—have been important only because of the volume of their use. General Motors and Ford virtually gave birth to laser disc as a business with their multi-thousand unit orders in the late 1970s, and the Department of Defense's EIDS training system procurement will be a major milestone in the evolution of disc technology. The sheer scope of the EIDS procurement—tens of thousands of IVD units, hundreds of millions of dollars—is destined to shape the near future of the industry. It will empower new players, define new technologies and possibly deliver some breakthrough software. Already it has thrust little Matrox, a Canadian hardware supplier, to the forefront of those who sell the "iron" (as the gear is termed).

However, through all of this activity (and hundreds of lesser efforts in industry) there have been few great achievements to date. It is true that you can see many models of cars on discs, and it is true that the operation and maintenance of the Army's Hawk missile system has been totally simulated on disc, but neither application is much more than adequate. Both are film-based and neither, therefore, has the aliveness and immediacy of video. They are, at best, rather boring visual museums of experience.

Videodisc is about video. It says, or should say, "live." Simulations should be just that—compellingly real and confusingly similar to reality.

Point-of-Purchase Displays

The realm of point-of-purchase (POP) displays is rife with disc products that may have been conceptually thrilling to the executives who spawned them, but which lost something in impact on the way to the kiosk. I have yet to see a POP disc that I really like. They seem to range from overblown exercises in Mirage special effects to banal assaults by teleprompter readers.

Nowhere can I find what Rob Lippincott (multimedia point man for Lotus Development Corp.) has termed a "user-obsessed" system. I do not want "friendliness" in these systems, neither do the executives who pay for them, nor do the users, who in the end they should delight. I want passion, and the user wants play.

One of the more adventurous POP systems yet undertaken, *Search and Source*, was the brainchild of an interior designer named Robert Sherman. Sherman's goal was to bring together, in a gigantic visual database, nearly all six million items of furniture, wall coverings, fabrics and paints used by interior designers. With this system, designers would be freed from the paper catalogs they have relied upon for a century. Price and availability information would be accessible online, and everything would be correlated according to a sophisticated color matching scheme. This system could have replaced rooms full of paper records and answered vital questions of availability before an entire design unraveled in the face of a grim price or intractable delivery realities.

Search and Source was conceived as a designer aid, but ultimately it could have been a point-of-purchase system worthy of the name. The product failed in its first run at IVD delivery because of shortages of capital and problems with the computer system that drove it. But *Search and Source* should be a beacon to every auto maker, computer company and cosmetics manufacturer who ever thinks of using POP systems. And this concept is coming back, in spades, on DVI (digital video interactive).

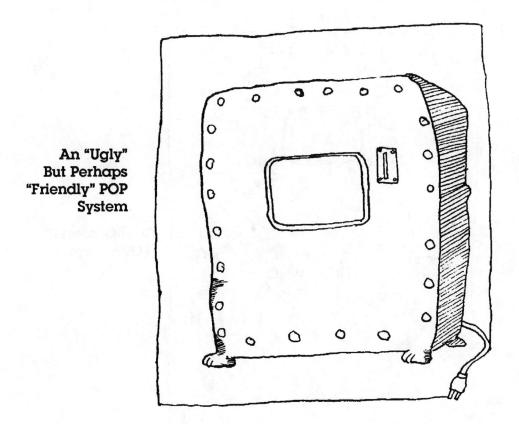

An "Ugly"
But Perhaps
"Friendly" POP
System

Empowerment of the User

Renault and GM could let people design their own cars. Apple and IBM could let customers see the operation of any of the software that runs on their machines. Revlon and Estee Lauder could present a gallery of faces and the cosmetics that enhance each; then customers could match their skin tone and type to the particular foundation and preparation that best suits them. The purpose of each of these hypothetical systems is to beckon the customer into the manufacturer's world.

The goal of interactive video is empowerment of the user. It should not be seen primarily as a vehicle for demonstrating the cleverness of media managers, nor should it be considered as a support for merchandising as we have known it. Interactive video is a new way to do business. It puts the customer or the learner at the heart of the process where he or she can make vital choices, not idle and trivial selections in a funky slide show.

Interaction brings viewers in. That is where you want your customer or trainee to

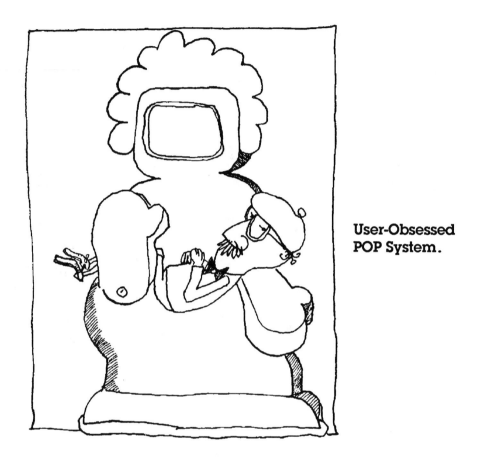

User-Obsessed
POP System.

be. Don't hold them off while you dazzle them with video footwork. Consider only this: What do they need?

Museum Exhibits

Discs have only recently been used in museum exhibits. The very first museum use of laser disc—a visitor information kiosk in the Wing of Primitive Art at the Metropolitan Museum of Art—opened in 1981. More recently, the *Nova Science Quiz* has appeared in several science centers. The *Golden Gate Disc*—a still-frame simulator ride over San Francisco Bay—is "wowing 'em" at San Francisco's Exploratorium. The Explorers Gallery at the National Geographic Society headquarters in Washington, D.C. features a disc-based geo-theater with a state-of-the-art interactive display projected around the six dozen lucky attendees. Some museums—the National Gallery of Art in Washington, DC, the Louvre in Paris and the Getty Museum in Malibu—have committed portions of their collections to disc for public consumption outside the museum. At this time, there are a few museum applications of disc, but this universe will grow dramatically. The Science Museum Consortium, based at Boston's Museum of Science, is pooling institutional resources to create interactive videodiscs on space, health and twenty-first century technology. Several projects are underway. Such collaborative efforts will have a positive impact on the distribution of this technology.

IVD is the perfect information directory tool, as well as an ideal archiving tool for fragile or dispersed collections, such as photo archives or drawings. And, discs represent a potential profit stream if sold through museum shops. Only on a videodisc, released from linear constraints, can one browse, pause and move about a museum in a fashion analogous to actually walking there.

Here, again, the challenge to producers will be to make the experience "live." We can use video, and clever interactive design, to offer powerful, varied and repeatable experiences to the viewer. The richness of disc is equal to these demands. The challenge is to our design capabilities and to our imaginations.

Trade Shows

Trade show use of disc has been virtually nonexistent. This is preposterous since discs are cheaper and more accurate tools for product display than distracted sales folk or large-scale commercial exhibits. Moreover, a computer-disc system can inquire after a user's precise application and model a product's performance in that setting. It can deliver specs, prices and delivery dates—and even provide a videocassette record of the relevant demo for qualified sales prospects. I believe that the videodisc can and will find a spot in the armory of marketers who use trade shows to sell their wares.

Customer support is another area ideally suited to the capabilities of videodisc. For tasks such as training a beginning PC user ("which way do I insert the diskette?"), the disc is an ideal delivery mechanism. Self-paced, private, patient and attentive, a well-designed IVD system can act as an "answer-man," when located in a store or sales office where it can relieve sales personnel and answer customers' questions. As a producer, you must take care to foresee the delivery environment and then you must produce to fit the situation. In 1989, Apple began rolling out a major customer support and training system on videodisc, with initial positive results.

Simulation Systems

Simulation systems for pilot training, driver training, surrogate travel and gaming are the high country of interactive video. A user may require seamless branching, which may well mean that there are two or more disc players driven by computer "ping-pong-ing" back and forth within the system. Or, systems may use interleaving and machines capable of instant track-jumping. Simulation environments are plastic realities. They are meant to impact the user with their verisimilitude. Once more, the producer must see beyond the daunting technical requirements of the task to the delivered product. It will not be enough to check off the blocks of the flowchart as roughly analogous footage is shot. The producer must begin at the end—foresee the delivery environment, feel the quality of the desired interaction, sense the interplay of sound and motion—and create that.

There are so many distractions in a high-level simulation project, so many layers of paper and so many issues of hardware coordination, that it is easy to take your eye off the doughnut and stare into the hole. The producer must coordinate the deluge of activities that will yield product—and look beyond. If the IVD producer fails to grasp, viscerally, what the product must be, who will be the keeper of that flame? The director may not see the whole picture; the editor will be too late; and the actors are only corks bobbing on this raging sea. Someone—the producer—must visualize the ultimate user's experience and shape everything to create it.

Games

Games once were—and may be again—the realm for creative stretching in interactive video. They foundered in a changing marketplace and on nonvisionary software. But experiments were conducted that promised brave new worlds of experience. The techniques for real-time interaction with movie materials are at hand, and the hardware to support such goals is around. But the arcades are currently dead (though faint intimations of life are detectable), and no new shape for experience within them was invented in the 1980s. In the home, MysteryDiscs are a receding memory and Clue/VCR, a poorly produced curiosity. It will be years before there are enough PC owners with disc players, or CD-I player owners, to constitute a market. Quo vadis?

In Japan, hundreds of thousands of disc players were sold to people who wanted to sing along with discs as Karioke took Nippon by storm. And, Pioneer built up a catalog of CPE (computer program encoded) discs that featured arcade games playable at home, via a MSX computer. If you missed Astron Belt at the arcade (as Bally no doubt wishes it had), here's your chance to play. Driven by the wind that is slowly rising in the successful Japanese market, we may see a reemergence of disc-based games.

CONCLUSION

For would-be producers of gameware, point-of-purchase discs, exhibits or simulation systems, the challenge and the need to stretch will be manifest. As a former actor, recently employed in government said, "You ain't seen nothin' yet." How true. There is work to be done.

3 How the Videodisc Works

Optical videodiscs contain information encoded as microscopic pits pressed into a spiral configuration on the disc surface. The information stored in these pits is read by a laser beam and transmitted to a decoder in the player. The use of laser light to read the signal means that the disc itself is not worn by use. If it is exposed to heat, it may warp and thereafter improperly reflect the laser, but the information encoded in its surface cannot be erased.

CONSTANT ANGULAR VELOCITY (CAV)

Constant angular velocity (CAV) is what we call the disc playback format used for interactivity. In CAV, each track, or 360° rotation of the disc, contains two video fields comprising a single video frame. The disc is rotated at a constant rate of 1800 revolutions per minute to yield 30 frames per second of play. Fifty-four thousand frames, or 30 minutes of linear material, can be stored on each side of a 12-inch CAV disc.

Individual still pictures can be selected from any of the 54,000 frames simply by repeating the same track on the disc again and again. On laser disc, a single frame can be displayed indefinitely without harming either the player or the disc. This capability to instantly access and then indefinitely display individual frames (FRAME SEARCH and STOP MOTION), perhaps more than any other feature, sets the videodisc apart from videotape playback systems.

By instructing the videodisc player to move from one freeze frame to the next, STEP MOTION is achieved. This can be done in either forward or reverse. STEP MOTION has many uses: one application could be to demonstrate in a series of stepped stills the disassembly of a device and then its reassembly; or one can decompose more

19

abstract problems, such as in mathematics or language arts, by presenting conceptual components as successive still images.

In SLOW MOTION, the videodisc player repeats a frame a specific number of times before moving on to the next. The user can set the rate, ranging from normal speed to STEP MOTION, in either forward or reverse.

In FAST MOTION, the player reads only one field of video (one-half of a frame) before moving on. This permits two or three times normal speed playback in either forward or reverse.

In SCAN MODE, the videodisc player skips several tracks at a time, resulting in extremely high-speed passage through the material. This allows rapid visual search, forward or back, at roughly 20-times normal speed. (No audio is currently available in any mode but NORMAL PLAY.)

Besides NORMAL PLAY and SCAN MODE, a user can randomly access a specific frame on a disc by number, using the remote control keypad supplied with the machine. Worst-case access time should be five seconds, or less. For example, the entire painting collection of the National Gallery of Art is available on a consumer disc. A user simply looks up painter and painting in an accompanying catalog and keys in the corresponding frame number; within a few seconds, the painting is displayed (for as long as the viewer desires).

Fast access to any frame or linear program segment means that program material can be read in any sequence, with little or no regard for the order in which the information was recorded. In an arcade game, this permits rapid switching from an "attract" sequence to actual game play; and in play, the results of a move can be quickly, sometimes instantly, displayed.

Every frame on a videodisc has an address or frame number, stored in the vertical interval (the space between frames), which is analogous to the numbers on the pages of a book.

Chapters can also be numbered on a videodisc. These numbers are stored in the invisible space between frames, so that entire segments of linear program material can be called up by chapter number. In the MysteryDiscs, viewers select their route through the mystery by designating a chapter number and audio channel. At the conclusion of the motion sequences, viewers are presented with a menu of still-frame clues that must be investigated to solve the mystery. The clues, and the correct solution, are arrived at by selecting frames via their address numbers.

Laser discs can accommodate up to 80 chapters on a side. No chapter can be shorter than 30 frames, or one second. If you use chapters on a disc side, then all the segments of that side must have chapter number codes. The insertion of these codes or cues can be done at certain commercial editing facilities or at the disc mastering plant itself, for a fee.

On a disc, it is possible to encode specific stills so that the player will stop automatically on a designated frame during play. This capability (which, be warned, not all videodisc players possess) allows a disc designer to integrate motion and still sequences without requiring user input to change modes. One use of this capability is to introduce chapter stop codes at logical points in the program (such as the moment when you instruct your trainee to turn to the workbook for drill and practice).

Most videodisc players have two sound tracks, though some new devices offer four. These can be used to provide high-quality stereo sound or to offer different audio information channels—in different languages, or at divergent levels of difficulty. On the MysteryDiscs, one finds a movie with two different sound tracks that offer different clues. A user's selection of audio channel 1 or 2 in four different chapters leads to 16 unique paths through the material, and the user must solve one out of 16 very different crimes.

The following list presents the basic mechanical capabilities of the CAV videodisc. These are the boundaries within which designers and producers of interactive programming must work. These constraints should be seen as challenges to our imaginations, since the permutations of these capabilities can yield endless excitement and memorable awakenings.

- NORMAL PLAY
- STOP MOTION
- STEP MOTION
- SLOW MOTION
- FAST MOTION
- SCAN MODE
- RANDOM ACCESS

CONSTANT LINEAR VELOCITY (CLV)

There is another videodisc player format—constant linear velocity or CLV. Here, equal length portions of disc tracks are scanned by the laser during any given interval of time. The play speed of the disc varies from 1800 rpm at the hub, where the radius is short, to 600 rpm at the outer edge, where the circumference increases. The result extends playing time to up to one hour on each disc side. However, the interactive options in this format are limited: on most disc players, it is no longer possible to isolate a single frame on a given track. CLV does *not* offer STEP MOTION, SLOW MOTION, FRAME SEARCHES or PICTURE STOPS.

Materials recorded in CLV can, however, be segmented into linear chapters. The disc player can swiftly search to any such chapter and begin play. For linear expository materials, CLV offers the advantage of accommodating more program length on a side. If a project requires only chapter-length slices of linear playback, CLV offers an expedient technique. One caveat: you cannot have both CLV and CAV materials pressed on the same side of the same disc. It is possible, however, to press CAV on one side and CLV material on the other.

Pioneer's LD-V8000 player allows the disc producer to store a frame and then execute a search, even in CLV mode. This capability means that on this high-end device, you can play linear motion chapters and still pictures in CLV mode, much the same as in CAV—with 60 minutes of material, not 30 minutes, on each side.

THE ELEMENTS OF INTERACTIVE DISC PROGRAMMING

Whatever the eventual application environment of a disc, the producer will select from a repertoire of functions those that best suit a project's needs. This repertoire includes still frames, linear video sequences, text and graphic overlays, and multiple sound tracks. The first job of the disc producer is to evaluate, within the measured constraints of budget and time, what range of interactive functions will best serve the communications goals of the project and will most intelligently utilize the capabilities of available playback equipment. In fact, this is a complex matrix of values, for there is a cost in dollars, time and attention to be weighed for each alternative.

To begin, one must know and appreciate the quality, value and power of each kind of software that can be retrieved from a disc or created by the interface of disc and computer.

The Still

First, there is the still. Utterly simple, but magnificent in its economy and impact. Each side of a 12-inch laser disc can accommodate 54,000 separate still images. Tools are available to encode sound with each still picture. Digital circuitry residing within specialized black boxes can pick off compressed audio stored in the video signal and in the vertical blanking interval. "Sound-over-still" requires special equipment to record and play back the audio. Sound is translated into digital signals which are recorded on the disc. Later, a decoder retranslates the digital pulses into an analog representation of the sound. Ordinarily, sound on disc is only available via playback at normal speed when it is read off the disc in real time. This is costly in terms of disc "real estate." The addition of frame buffers to top-end machines means that IVD producers can now load (and display) a still, and retrieve associated sound from one of four audio tracks stored elsewhere on the disc.

The still image, whether for text (instructions, explanations, background data, test materials) or for pictures (e.g., VPI's National Gallery of Art disc), offers a powerful means of condensing textual data or presenting pictorial databases.

Motion Sequences and Linear Program Materials

Motion sequences or linear program materials often carry the bulk of the message on a disc. Motion segments can lure customers to your kiosk, invite players to your game, describe the contents of a point-of-purchase (POP) disc, tour a plant, train your colleagues, explore dynamic processes or illustrate concepts. Applications are as varied as communications goals. These motion segments can be clustered together on the disc, or interspersed with still frames at either end or throughout. Remember that a motion sequence can be made to freeze at the end of a chapter (in a consumer disc) or at any point in a Level 2 or Level 3 system. When the motion picture freezes, graphics (including menus) can be overlaid on the screen in computer-driven settings.

Graphic and Text Overlays

Graphic and text overlays are an integral part of many videodisc systems. The choices of layout and information sequence are dictated by the video imagery stored on the disc, as well as the data structure defined by the flowchart. These elements of IVD productions are a striking example of the fusion of computer and video technologies in the realm of teleproduction.

It is also important to remember that the videodisc is the ultimate in non-volatile memory. It cannot be erased. In the event of nuclear war, our audio/visual bequest to spacefaring anthropologists could be our discs—they alone amongst our films and tapes will not be erased or readily burned (though they might warp). In line with this extraordinary durability, do not place any readily changeable information on your disc: prices, delivery schedules, store locations, model numbers and the like belong on computer diskettes for which the cost and effort of a change is marginal. For example, when creating a pioneering disc series for interior designers, a consortium of interior designers took pains to assure users, vendors and manufacturers that price and availability information would be continually updated and overlaid atop corresponding product imagery.

Multiple Sound Tracks

Multiple (two or more) sound tracks, recorded either on location or in post-production, offer several benefits to producers. Once their timing has been determined (with care), the production of sound tracks is straightforward. The playback unit has a simple switch to "flop" from one sound track to another, permitting tremendous flexibility in training or entertainment applications. The normal playback mode can instantly switch to a remedial or hint mode, depending on user behavior. The capacity to present one sound track and then another allows the user to "re-access" previously displayed video with entirely new meaning attached. One novel use of this technique appears on the art disc, *Lorna*, produced by Lynn Herschman. In one segment of this interactive narrative, the title character carries on a phone conversation either with her psychiatrist or with a travel agent, depending on whether sound track A or sound track B is in play. In the MysteryDiscs, different clues reside on different sound tracks. The two-volume

Anthology of American VideoArt has Japanese on sound track 1 and English on sound track 2. Dual sound tracks, when properly used, can more than double the information-delivery capacity of a disc. And, new digital technologies will multiply the number of sound tracks on a disc—to 16, and beyond. The Pioneer LD-V8000 state-of-the-art machine will detect and play back four channels, two analog audio channels and two digital audio channels, encoded in the video signal.

CONCLUSION

To recap, the following list provides a partial menu of IVD capabilities:

- Still frames (pictures and/or text)
- Motion sequences (linear video segments)
- Access to individual frames
- Access to individual linear segments
- Step motion through a series of still frames
- Automatic picture stops
- Automatic chapter stops
- Multiple sound tracks
- Slow motion forward or reverse play
- Fast motion forward or reverse play
- Graphic overlays
- Integrated display systems—whose keyboards and touch screens play a decisive interactive control role.

The list below summarizes the structural distinctions between "normal" linear teleproduction and producing IVD.

- Discs use still frames.
- Branching is basic to disc.
- Access or search time requirements must be considered on IVD.
- Disc real estate considerations are unique.
- Textual or graphic overlay capabilities alter display strategies.
- Utilization of multiple sound tracks can provide different information.
- The possibility of "re-accessing" or reusing video, with different sound tracks, multiplies the uses of pictures.
- Integration of computer and video imagery creates a new medium.

Keep these expanded boundaries in mind when discussing the possibilities of disc with your clients. In the following chapter, we will discuss the seven phases of IVD production and how they can be used to develop a realistic budget proposal for your disc .

Part 2
Production Techniques

4 The Seven Phases of IVD Production and Budgeting

In videotape production, there are three standard phases: pre-production, production and post-production. Pre-production entails the planning of activities; production refers to the actual recording of events; and post-production is the editing or assembling of the final product. With IVD, however, seven different phases of production can be identified. These are:

1. Project development
2. Design
3. Pre-production
4. Production
5. Post-production
6. Programming
7. Validation.

I have seen slightly different lists of seven, and sometimes eight, key steps of videodisc production (adding, for example, "system integration"). No such list is "holy." It is just a typology—a way of sorting out distinctive activities. In the following pages I will define these seven steps and show how, in a simple formula, they can be used to out-line a realistic budget proposal for any interactive videodisc project.

THE SEVEN PHASES OF IVD PRODUCTION

Project Development

The development phase (sometimes called the proposal phase) of interactive videodisc production entails the thoughtful translation of a client's project goals into a preliminary design for an interactive product. The type and level of interactivity is defined, the delivery system is specified and the general treatment of content is decided upon—and all of this suggests a budgetary outline. In other words, here we translate the

27

client's itch for interactivity into a product definition, project outline, tentative schedule and approximate budget.

A good bit of specialist knowledge must be focused at this stage to assure that a reasonable approach to the entire project will follow. So, wherein linear production producers may be expected to "spec" and bid a project for no compensation, a rather substantial front-end investment is required to define an interactive project. It is not unreasonable to bill 10% of the entire estimated project budget for these design and development services.

On the client side, this implies a fair amount of homework and internal support for an IVD product prior to bid solicitation. It is the client's job to know—or to quickly learn—the differences between the costs and capabilities of CBT (computer-based training) and IVD. I think it is unfair to expect disc producers to provide the equivalent of a graduate training seminar on interactive technologies before a client is prepared to declare him or herself a qualified candidate for disc use.

Producers, beware the enthusiasms of the ignorant. I once, to my regret, played disc mavin to a production company more "in love" with IVD technology than in-the-know. We wasted copious amounts of energy dancing on the desks of a major bank trying to sell discs to train money managers in the use of a mainframe computer system. It should

"See yourself, and help your client see you, clearly. As an IVD producer, you bring specialist skills to particular and definable problems addressed in the project development phase."

have been obvious from the start that it is wisest to use computers to train employees in computer usage—it is also cheaper. How and why disc specialists got involved in disqualifying this clearly inappropriate client, I'll never know. It should have cost someone money; as it went, it just burned some time.

There are some decent sources of information toward which you can point your would-be client. Remember those other relatively inexpensive training tools: the printed page and computer-text screen.

The Videodisc Monitor, published by Future Systems, is the most complete monthly intelligence report in the field. Its back issues are a road map through the hardware wars and applications investigations of the past decade of interactive video and IVD. Vidmar Communications publishes the *Optical and Magnetic Report* with excellent inside industry information. Pioneer, 3M and Sony offer fairly substantial booklets on the several steps involved in preparing and replicating videodiscs; these are available free upon request. There are several trade magazines that publish entire issues, each year, dedicated to interactive production. *AV Video* (Montage Publishing, Inc., White Plains, NY) covers interactive trends on a monthly basis. The International Interactive Communications Society (IICS) is the leading organization of interactive producing professionals worldwide. Its local meetings and newsletters are vital information resources. Demos are frequent and extremely valuable. This is where you can meet your peers and see their work. The International Television Association (ITVA) is an organization of video professionals in the business and institutional TV fields, which has a Special Interest Group (SIG) for Interactive Video Producers. Meckler Publishing has a trade show in IVD, as does the Society for Applied Learning Technologies (SALT). Knowledge Industry Publications, Inc. presents video and multimedia expos that offer a lot on IVD. A new trade association, the Interactive Video Industry Association (IVIA) has been established recently with its showcase at Tech 2000. These, and other sources, provide both introductions to the technology and updates on recent developments. It's not a bad idea to point a client toward these sources and await a considered request for proposal.

Video is great for fostering perceptual readiness, for simulations, for conferencing, for role-modeling complex behavior and for demonstrating dynamic processes. If the training task goes beyond this to specific and precise technical behavior, consider alternative media.

Design

The design phase of a disc production project is of crucial importance. Here, the content is translated into known numbers of motion picture, still frame and computer text, or graphic frames. The routing through the interactive branch points is determined, as are test sequences and menu locations. The frequency and location of remedial branches is set forth. The complete workings of the interactive system are decided upon, committed to paper and distributed to all members of the production team.

The document that contains all of the above data in its densest form is the flowchart. The flowchart is a graphic notation that evokes the relationship of each part to the whole, and offers a shorthand sketch of where and how users interact with the content. Once the disc design is formalized and agreed upon, the realities of production can be addressed.

Pre-production

Pre-production for an IVD project is similar to the pre-production phase of linear videotape production. During this phase, the production crew is assembled, facilities and locations are engaged, the cast is selected and scheduled, all of the logistical support for a production is prepared, and the script is finalized.

Production

Production is the phase during which audio/visual materials are actually created and recorded. In a studio or on location, with actors or company personnel, using real images or computergraphic animation, the content of the script is turned into motion sequences and still pictures that are recorded on tape or film. For IVD, timing and freeze-frame requirements, as well as attention to branching storylines, subtly alter the normal tasks of teleproduction.

Post-production

Post-production means editing. The disc post-production phase involves an extremely precise and often arduous process. The recorded material is assembled with constant consideration being given to relative locations of sequences and overall use of available real estate, or time, on the disc. Still frames, which are sometimes scattered throughout the disc, must be included in the assembled product.

Edits

There are two phases to all video edits: offline and online. Usually, it pays to create a rough version of the product on a relatively inexpensive editing system. This gear is, in every sense, off-the-(expensive)-production-line, hence the name. In recent times, the standard has been to use 3/4-inch U-matic editing equipment. The essential product of the offline system is the edit decision list (EDL). The EDL notes the start and stop times of every motion segment to be included in the final product. After review and approval of this "rough cut," the producer takes the original master material and cuts it at an expensive online facility utilizing the time-coded edit decision list as the road map. Stills may be tested in a rough cut but are rarely fully executed there. Stills are usually added at the end of the post-production process, during the "fine cut."

The Pre-master

The final product of the video editing process, when aimed at disc replication, is a "pre-master" (called this because it is the last artifact before the disc master pressing template itself). There is nothing mystical about a pre-master. It is simply a 1-inch, D2, 3/4-inch or Betacam SP videotape from which the physical disc master mold will be created. This tape may or may not contain information that causes machines to come to an automatic stop at the end of a "chapter," or which causes computer codes recorded on the tape to be dumped into the internal microprocessor of a Level-2 player. And, if intended for a Level-3 application, the pre-master may be difficult or impossible to distinguish from any other master tape ever created, except that it probably contains a number of stills. Pre-mastering simply means getting a final tape ready to submit to the disc replicating company.

Programming

Programming is the writing of computer instructions that will operate a disc in a Level 3 or externally computer-controlled application. Programming can take time. It is important to begin this activity before disc replication so that bugs can be eliminated from the working system. Depending on the scope of the project, disc programming can range from a straightforward two-day task to a memorably horrendous multi-month bummer. It can also take two hours with *HyperCard*. In general, it is wise to hire consultants early. Assess the project's requirements accurately, and get to work as soon as the flowchart is finalized. The best way to find a consultant is to ask for a recommendation from someone who has done a disc before. Try to avoid first-timers; they can be costly when it comes to the one item you can't buy later—time. Good computer-folk charge about $35 to $50 per hour for their work.

Validation

Validation is the process of assessing the performance of your system's users, or of tracking the pattern of utilization of materials on your disc. It is often necessary to validate the usefulness of this complex and expensive technology by demonstrating enhanced performance by users (vs. non-users) in training or sales applications. Testing behavior before and after exposure to interactive training programs or sales aids can be invaluable, particularly when marketing future systems to a client.

A computer-managed (Level 3) interactive system can include testing and tracking functions that will measure the pace and reliability of skill acquisition by disc users. This feature has more than cosmetic appeal. It can be used to certify trainees. It can also be used by manufacturers or service providers to reduce insurance claims by providing solid evidence of proper personnel training. For example, a training disc aimed at bartenders attempts to describe the subtle signs of alcohol intoxication. The intention is to educate bartenders so that inebriated persons will not be served—lest the bartender and bar be sued under various Dram Shop Liability laws. Use of this training disc and associated performance tracking software may help lower Dram Shop insurance premiums, and may be introduced as courtroom evidence of good faith by the service provider.

TIME CONSIDERATIONS

Now that we have identified the elements of IVD production, let's look at the time frame involved and detail scheduling considerations.

The development phase can consume three to four weeks of time as project managers at both the client and producer ends familiarize themselves with the project's challenge and its interactive solution. This is management time; it is expensive.

The design phase can take four to six weeks to complete. Each subsection of this work—instructional design, IVD design and flowcharting—can easily consume two weeks. These elements can overlap somewhat, and each completed design component must be reviewed by the other players. Project management must allow ample time for revision and team consultation.

Scripting can take four to eight weeks, depending on the scope of the work. A small project may of course require less time—but all scripts require extensive review. Ample time should be budgeted for revision, recirculation and approval of evolving script drafts. In the project contract, explicit attention should be given to this matter. The client must agree to a preset schedule for review and turnaround of each script draft.

Initially, the scriptwriter, or writing team, prepares an outline of content keyed to the IVD design (one week). This detailed outline may be reviewed by the project team or producer. A first draft of the script is then executed (allow two to three weeks). The client must be allowed at least one week for circulation and revision of this draft. Thereafter, an updated draft can be prepared, which incorporates the changes (one week). Final review, revision, and ultimate sign-off on the script can consume two weeks.

Pre-production requires two to three weeks for hiring actors and crew, engaging locations or studio, arranging insurance, catering, transport and lodging. This period of pre-production brings the producing team together and up to speed. There is ordinarily no interval between pre-production and production. The director must be intimately familiar with the work to be done, the crew must be finely tuned, and the actors must have digested the script, before this period closes.

Production is usually a two- to three-week phenomenon. Shooting may take three days only , but load-in and load-out of the location, props return and final accounting can easily span two weeks. Two weeks of actual production (i.e., shooting) days bespeaks a considerable investment of energy and money. The investment can be mitigated through ownership of studio and/or location shooting facilities, but production is production—it means crews, actors, props, call times, logistical coordination, equipment, food, petty cash, and lots more. Even on the smallest scale, this means money. A production day with a crew of five in a corporate-owned facility, shooting two actors with two cameras can cost $3,500, not counting tape, lunch, transportation or hotel rooms, and assuming that the company owns the recording equipment and is already insured.

The point is: shooting days are costly. Budget the minimum feasible number of days, hire a knowledgeable producer, and have a tough "budget watcher" with good interpersonal skills as a production manager. No one—not IBM, not AT&T—can afford to plan four-week shoots. Four weeks is for a movie; we're talking disc.

Post-production's time-consuming phase is the rough cut. Allow at least a week for the first assembly of the product. Basically, allow at least two and a half editing days for every day you shoot. Again, as in scripting, allow for review by the client, and build that time into the contract. After the rough cut approval comes the final or "online" edit. Here, the per-hour costs can be staggering. As in production, no one can afford to be expansive in budgeting time for this work. (In a New York online suite, I heard the editor joke, "The goal here is to *leave!*") Plan for a plausible minimum and give yourself a 10% to 15% contingency. One week of online is an awful lot of fine cut time. Online is machine time. The goal is to execute the final product as fast and as surely as possible, conforming with the frame numbers already determined in the rough cut. The only creative work done online is special effects, text and still frame generation, which usually cannot be done offline. Here, creative time must be allowed. But usually these elements of the project are highly constrained, and a competent facility can give you a reliable estimate of the time that will be required.

Online post-production time can be guesstimated at the rate of roughly six edits (cuts, not dissolves) per hour. You may do better, but don't bet your life on it. One day is a low-end limit for an 80 to 100 cut disc (a simple one); two to three days is more typical; five days is for big projects.

Programming, for Level 3 projects, may begin prior to or during production. Control software can be designed as soon as the flowchart is finalized. Executing and debugging a custom computer program takes time, so the programming phase should straddle the shooting and editing activities to prevent delay. Optimally, the debugged program will emerge just as the online video edit is completed. The two products can then be melded together and the result tested.

Testing and validation can be as brief or extensive as budget and project goals dictate. Some proof of performance is required in any case. Training and point-of-purchase products must be fully tested before release. So, depending on your output, this phase can take as little as a week or as much as two months (or more). A tape-based emulator allows pre-pressing check-out of a disc's interactive functioning. So, of course, will an OMDR (optical memory disc recorder) disc. Inexpensive test pressings permit performance evaluation of all features of an interactive disc. At $600, or considerably less, one-disc pressings are available from several sources. Crawford Communications in Atlanta, for example, will do one-offs for $300. So will Skywalker Sound's EditDroid subsidiary in Los Angeles, CA. The best course before final replicate pressing may be to test one disc fully. Validation of the efficiency, impact and appeal of an interactive product, as well as its technical functionality, may be a part of this phase.

PUTTING IT ALL TOGETHER IN A BUDGET PROPOSAL

There are, of course, an infinity of possible budgets with greater or lesser profit margins, higher and lower production values, to match an infinitude of creative challenges. Each producer approaches the budget challenge resonating to his or her entire accumulated experience, expressive goals and pecuniary desires. There is, therefore, no simple formula for the "right" approach to a budget.

What I offer below is a synthesis of my own experience. It has informational value to me because it reflects actual numbers that I have worked with. It also reflects a mental construct that is an amalgam of experience, desire and game theory (guesses about what my clients surmise about my goals, and about what I will agree to do). Use it with caution—it does have some predictive value, but your production will be unique and will benefit from the consultation of a good production manager before you submit a final budget.

Four Budgetary Phases

We can divide the video production phases of an IVD project into the following four budgetary chunks:

• Development (includes proposal and design work)

"Remember that the interactive videodisc production process has numerous pitfalls—all of them expensive. Your budget is a tool for prediction and control."

- Pre-production
- Production
- Post-production.

Overall, rough and ready, here is how I expect the money to break down as it will be allocated to each of the four phases:

- Development   10%
- Pre-production   10%
- Production   55%
- Post-production   25%

I recommend a 10% contingency against the unforeseen, extracted proportionately from each phase.

The Interactive Specialist Elements of a Level 3 Project

The interactive specialist elements (activities unique to IVD) of a Level 3 disc project include IVD design, still frame production and computer programming. These elements can consume 30% to 50% (or more) of the project total. A reduction of this percentage can be derived from proportionate reduction of the pre-production, production and post-production expenses of the other (expository or narrative—i.e., linear) materials.

A breakdown of the linear production figures permits a line-by-line analysis. (See Figure 4.1.) Again, these are rough and ready figures that I have extracted and averaged from dozens of projects. Your numbers will vary.

Figure 4.1: Breakdown of Linear Production Figure

Design and Development:	
Script	6%
Producer's fee	4%
Pre-production:	
Art dept. (props)	5%
Production management	5%
Production:	
Art	7%
Director	5%
Facilities	15%
Crew	10%
Talent	8%
General expenses	10%
Post-production:	
Offline edit	5%
Online edit	12%
Special Effects	5%
Audio, voice-over, music	3%

There is no limit to your interactive specialist investment in a project. It will be shaped by your intended delivery environment, number of still frames, number and complexity of branches, amount of overlaid text and graphics, and the complexity of your computer control program. Your only way to control these expenses is to carefully define your design, as early in a project as possible, and work up a budget for each proposed activity.

Other Cost-Intensive Variables

Clearly, there are numerous variables that can act as cost drivers during the production of an IVD's linear elements—other than weather, accidents and illness. These drivers, which can impact your planning, include:

- Cast size
- Number of sets
- Number of locations
- Studio size
- Number of cameras
- Choice of tape format and recording equipment
- Number of still frames
- Number of branches (and number of scenes)
- Quantity of special video effects
- Producer's and director's fees

Simply put, the larger the number of sets and settings, the larger the amount of time spent in re-lighting and preparing to shoot. The smallest false move on a set can easily consume an hour before tape rolls again. A change of location entails build-time, propping, trucking, technical checkout and general set-up delays, which can take a day. The economics of such delays can be decisively adverse if they are not properly foreseen, and managed.

CHARGING YOUR CLIENT

A companion question to "How do we budget?" is "How much do we charge?" In response to the question "How much is a production? Just give me a ballpark," I've heard the response, "How much is a house? Just give me a ballpark." There are many right answers. In video, you can do anything for any price—the "thing" just changes with the price. Rules of thumb are hard to come by. One sage has remarked, "You can have it quick, good or cheap—pick two."

First, crystallize your production plans so that you know the number of sets, size of the cast, and number of planned days of shooting and editing. You can engage a production manager—if a script already exists—and come to a very precise estimate of these costs. But in the opening stage of the project, probably only bare outlines of the production's scope will be apparent. How do you close the deal—get the client to commit—if

Figure 4.2: The IVD Budget Formula

Let $\sum r$ = Total real estate
          or running time (in mins.)

   S = Total number of script pages

   S/5 = Total number of shooting days
          (at 5 pages per day)

$\sum r/4.5$ = 1 online edit day—yielding 4.5
          minutes of pre-master

Assumptions:
     $8,000 per shoot day
       $750 per offline edit day
     $3,000 per online edit day

Total
Production $\text{T\$} = \boxed{\text{S/5 x 8000}} + \boxed{\text{2.5 S/5 x 750}} + \boxed{\sum r/4.5 \text{ x } 3000}$
Cost

                       shoot   offline   online
                                   (Recall: 2.5 days
                                   of offline for
                                   each shooting day)

Programming:
     $n_1L + n_1I = \sum r$ n_1L = Number of linear sequences x their
                                              running length
                                  I = Interactive text or still frames

     $PT = \partial(P \text{ x } 400 + E \text{ x } 100)$

Total
Programming   = ∂ (a driver) represents the amount or frequency
Cost   of interactive branches and/or stills
                        P = Programmer days @ $400/day
                        E = Software engineering or high-level
                              technical coordination @ $100/hour

---

you cannot tell them, roughly, what the product is going to cost? Even before the development phase commences, your client will want to hear a number, or a range. By applying a crude but effective formula, you can know what your costs will be.

The Formula

Step one: slice the project into interactive elements and linear production elements, and assess the scope of programming time and hardware integration that will be required to make the delivery system work. Multiply the number of programmer days by your cost (probably about $400 per day) and add engineering time at $100 per hour.

Step two: estimate your production expenses. Assuming that you have a single location, one shooting day per five (or six) script pages, a crew of 18 and a cast of five, a single-camera shoot ("film-style"), recorded on Betacam SP should cost about $8,000 per day (you might call it $10,000).

Step three: post-production costs can be guesstimated. Offline: 2.5 edit days per shoot day at $750 per day. Online: assume a yield of 4.5 minutes of pre-master product per edit day at $3,000 per day. Alternatively, assume 100 completed edit events per day. If stills are a key component, in a non-archive facility, assume a maximum yield of 200 text or still frames per 10-hour day. (See Figure 4.2.)

Adding It All Up

According to our figures, a five-day shoot with a cast of five is a $50,000 event. It is followed by a $6,000 to $9,000 offline edit session and, for a 20-minute product, a $12,000 online edit. It is possible to execute such a piece of work for less, but these numbers will let you breathe.

On top of these costs, you must add overhead: assume that this project will tie you up for 8 to 12 weeks. How much of your office and administrative expense during that time belongs to this project? Add that in.

PRICING STILL FRAMES

How does one price still frames? Beware the facile promise to your client. Beware your hopes that it will be easy. Remember that you may have to check each pictorial still frame four, five or even six times by hand—at $250 or more per hour. And recall that each text frame entails many steps—design, pre-build, lay-in, error checking—each with a cost. Moreover, placement of the still frames themselves can require significant investment of time and other resources. A few years ago, I would have laughed at what I now use as a rule of thumb for budgeting still frames that are to be frame-accurately sprinkled throughout a disc—$50 apiece, as a rough estimate. There are frequent occasions when it will only run $10 to produce the frame, or even a dollar. But, it is simply dizzying to count the number of occasions on which one must manually check the pre-master to be certain of a frame's location if that address is frame-specific (for example, when the disc is tied to an instructor's guide, with page numbers exactly corresponding to frame numbers).

PROFIT

There are two ways to figure profit. Profit can be "clean" profit, beyond your general and administrative expenses plus overhead, in which case 15% is a reasonable target figure. Or, profit can be everything beyond direct costs. Most small companies cannot sort out and apportion project overheads, so it is easiest to lump overhead and profit together, tighten one's belt and hope for the best. The magic number for gross margin in this type of situation is 35%. You can aim for 40%. If you try to do better, you may not be competitive, or you will have to be sensational at containing your direct production expense.

No one gets rich doing training discs or POP discs or today's entertainment videodiscs. The trick is to stay alive.

"Technology is explicitness," McLuhan observed. Planning and budgeting an interactive videodisc requires utmost explicitness. You must strain to define, at the finest level of granularity, what tasks you face and how you will address those challenges. That discipline is what budgeting is about.

5 The Design Dimension

Once you have developed a budget and received the green-light for your interactive disc project, it is time to begin formal disc design. In this chapter, we will define the steps in videodisc design and consider the documentation that usually precedes disc production.

DELIVERY SYSTEMS

Given the broad range of capabilities of videodisc players and the fact that these devices can be teamed up with other machines, such as computers, a very wide spectrum of interactive experiences can be offered to users.

One of the first steps in videodisc design and production is to define the setting within which the disc will be used—that is, to select a delivery system. This choice will help to define many other elements of your task.

During the early 1980s, to help videodisc designers and users typologize the range of experience available on disc, the University of Nebraska Videodisc Design/Production Group defined the following discrete "levels" of videodisc interactivity:

- **Level 0**—no interaction implied or intended, such as when you turn on the disc player and cross the room to watch a movie.
- **Level 1**—videodiscs depend, for their control, on the intelligence of the user. They usually have motion video sequences and often have still frames; frequently they will have chapter stop code information as well. These discs are usually played on consumer-grade players and respond to commands input through the controls on the player itself, or via its remote control keypad.
- **Level 2**—videodiscs carry some control instructions on the disc itself. As soon

41

as the disc is brought up to play speed, it downloads this information into the microprocessor that resides in Level 2 industrial disc players. Here we can say that intelligence (instructions for use) resides within the playback machine. Level 2 discs are frequently found in point-of-purchase applications, since all that is required to create an interactive kiosk is the semi-intelligent disc player and a display monitor. An example of a Level 2 application is the J.P. Stevens white-sale disc, wherein the frame address of the sheets' sequence is loaded into the player's memory. A user need only touch a single button indicating sheets and the player races off to retrieve that segment. Otherwise, the user would have to key in a five-digit string of numbers and a "Search" command to arrive at a chosen location.

- **Level 3**—refers to systems in which the intelligence is completely external to the disc player and resides in an outboard computer. Such computer-driven videodisc systems offer the maximum in flexibility and user-responsiveness. Here, one can combine computer-derived text and graphics with disc-derived pictures; one can track user performance; and one can flexibly re-program the workings of the system. This configuration offers the optimum in self-paced instruction and user-sensitive testing and tracking. In Level 3, the computer can be seen as an elaborate controller for the videodisc player and the disc player can be seen as a computer peripheral with the disc itself as a visual database. Some people apply the term "video computing" to this mix.

- **Level 4**—refers to an integration of analog video and digital data capabilities. Level 4 videodisc systems will have massive text and pictorial storage and sound-over-still capabilities. This term has only recently begun to appear in IVD literature and at this date embodies more an aspiration than a product line. According to recent Nebraska materials, Level 4 refers to integrated videodisc-computer systems. Videodiscs in such a system contain analog audio-video and machine readable digital code. The videodisc player functions as an optical storage drive for the computer, and the computer is programmed from the digital code on the videodisc. Several senior IVD designers resist this new addition to the interactive typology. The first three levels are universally accepted as standard, and Level 3 has long been understood to include everything that involves treating the disc player as a computer peripheral. So, one might argue that Level 4 is a subset of Level 3.

Delivery systems are the configurations of hardware that utilize and realize these various capabilities. Everything that you might do as a videodisc producer or project manager will be defined by the choice of the ultimate destination of the disc—the delivery environment. If you are designing and producing for a Level 1 system (consisting of disc player, remote controller and monitor), you will have to decide how to slice the program into chapters and where to locate any stills you may use. For a Level 2 system, you will probably alternate short sets of stills (through which users will make selections) with relatively brief linear segments. If the goal is a Level 3 product—as is increasingly customary for educational uses, and always required for gameware and simulation systems—you will probably divide the display tasks between the disc and the computer, using the latter for graphic and text screens, as well as for partial overlays of video sequences.

A critical dimension of the delivery system pertains to user input. In a Level 1 system, you are limited to the hand-held remote controller or the on-board controls of the videodisc player. For a Level 2 or Level 3 system, the choices are broader: touch screens, keyboards or specialized control panels are options. The user can designate in numerous ways their next destination, and the system will respond. In Level 3 systems, the control variables are still more numerous. Military training and flight control systems of the future will tie videodisc databases to helmet mounted stereoscopic displays. A space station repair person will only have to look at a particular control panel to have a full-color display of its innards projected on the helmet's visor. A spoken command will suffice and the target device will be shown disassembled, or in any intermediate state that might aid the technician in repair work.

From a designer's or producer's perspective, the important point here is that the final destination and use of the videodisc is crucial in determining which creative tasks he or she will undertake. You must know where your disc will be played to create something useful. The range of delivery systems is broad, though the Level 1 to 3 typology is quite elegant and inclusive.

To create a working videodisc you must begin, conceptually, at the end of the process. Exactly how will the disc be played? Precisely what controls will be available to your user? What information, if any, would your system require about the user's needs? The answers to these and related questions, defined by the delivery system, will be embodied in your program. Figure 5.1 provides a checklist of hardware considerations for a delivery system.

Figure 5.1: Checklist of Hardware Considerations for the Delivery System

Cost
Memory
Graphics capability
Text
Overlay capability
Mode of access and control (keyboard, keypad, touch screen, etc.)
Help capabilities
Special features
Documentation and system maintenance

DOCUMENTATION

Once you have decided on your delivery system, you begin compiling the information you will need to get your IVD project ready for production.

Much can be said about the need for proper and up-to-date documentation for videodisc production. The essence of the matter is that you, as producer or project manager, must have the right stuff at the right moment—or your entire project can be imperilled. I witnessed a giant corporation pump out prodigious quantities of paper, yet it was never up-to-date with the actual needs of the production. Box loads of multi-hundred page scripts were frequently delivered, yet we had to resort to cut-and-paste jobs, which were swiftly photocopied on the set, so we could have what we needed.

As in every other aspect of disc production, good intentions and assiduous effort are only a part of the game. To successfully run this complex enterprise, you must be properly supported and documented.

"You Must Have the Right Stuff at the Right Moment, or Your Project Will be Imperilled."

The entire scriptwriting and production process is based on the contents of a design document, which details the needs and objectives of the program. Each objective is tied to a particular lesson or narrative sequence. This master document is a schematic model for your complete work. It is only the scaffold, however, because it does not describe in detail how particular objectives are to be achieved.

THE DESIGN DOCUMENT

The compilation of a detailed design document is a critical step in disc design. This is often a loose-leaf notebook that incorporates input from an instructional designer, an interactive videodisc design specialist and subject-matter experts. It begins with a front-end analysis that details the training needs of the audience and/or the learning and behavioral objectives of the program. After these elements have been discerned, the design document is passed on to the instructional designer. At this stage, the outline of the structure of the program is tied to the scope and sequence of lessons to be included. Each lesson is usually assigned its own page(s) in the notebook, which corresponds to a particular learning or behavioral objective. This objective can be noted at the top of the page. The content and visual treatment of the lesson is then briefly outlined. User

"To Successfully Run This Complex Enterprise, You Must be Properly Supported and Documented."

activities, including tests, are noted as well as branching outcomes in which each user response must be detailed. This material is provided by the IVD designer, who is careful to describe the overall architecture of the disc, citing how users may, for example, branch to remedial sequences and then return to the main body of materials.

Ordinarily, the instructional designer on a given project prepares the design document. Just as often, however, the disc producer doubles in this role. In either case, in its final form, the design document should answer these questions:

- What are the learning or behavioral objectives of the disc?
- What does the audience need or want?
- What is the instructional sequence?
- How will the hardware operate? Does it use a keyboard, a touch screen, or other user interface?
- How can instructional strategy make the best use of the available hardware?
- How can menus, help functions, indexes and glossaries best be used on the disc?
- What is the optimal media mix? How many still frames, how much motion, and how frequent should tests be?

From this material, a disc flowchart is prepared which provides an overview of the entire program's function. The flowchart provides useful material to the disc designers, programmers and instructional folk alike.

THE FLOWCHART

The flowchart tells scriptwriters, production personnel and computer programmers what the order and relationship of elements on the disc will be.

Flowcharts are graphic means of describing a sequence of operations performed on data. Flowcharts use a two-dimensional pictorial format, combined with some limited wording, to convey in explicit but shorthand fashion the structure and logic of a program's operation. There is a lack of uniformity in nomenclature and symbol usage in this realm. Sometimes flowcharts are called logic diagrams, process charts, block diagrams or system charts. There are several levels of detail that a flowchart may address. Typically, in disc applications, the unit of detail is an operation or short sequence of operations—as compared with the unit of detail in a system chart which is usually the work done by an entire subroutine or program. Flowcharts emphasize how data is transformed.

The flowchart provides a way of communicating, from one person to the next, the nature of the operation to be performed and of the data upon which it is to be performed. They are used by programmers as the basis for writing programs, and are vital as a means of communication among the team members who are participating in program development. Since they are pictorial and economical of detail, they are powerful and efficient tools for studying the structure of an interactive program. Flowcharts tell a great deal at a glance.

Figure 5.2: Interactive Flowchart Symbols

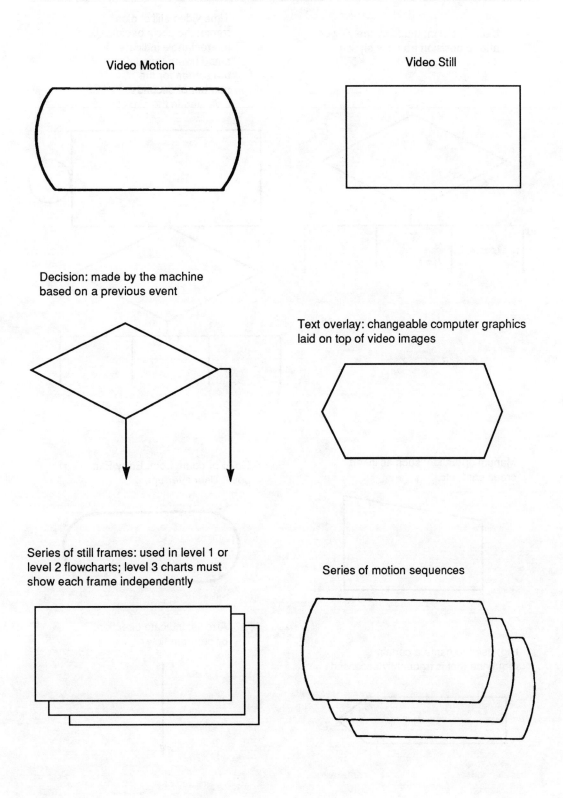

Figure 5.2: Interactive Flowchart Symbols (Cont'd.)

User Decision: made by the viewer
after a decision frame is shown

Time video still choice
frame: the circle beside
the rectangle indicates a
timed frame (one left on
the screen for the
number of seconds
indicated in the circle)

Manual operation: such as insert
credit card, etc.

Go to or come from. Entry/Exit
point. User interrupt.

Arrows: indicate direction
or program flow

Tag: used to name a certain
sequence that is frequently accessed

There is an extensive array of general and specialized outlines or symbols used in flowcharting. Many of these have been simply borrowed from computer programming. Others are specific to video applications. The symbols presented in Figure 5.2 are most frequently encountered in interactive videodisc applications. The wording within the outline defines the operation and may also describe the data involved. Typically, short English phrases are used.

It is common for a flowchart to evolve through the various stages of a project. It may begin as a block diagram describing the overall flow of the program and grow in precision and level of detail until it completely describes the disc users' possible experience(s), step-by-step.

In any case, the flowchart's power derives from its elegance and economy. By taking full advantage of the flowchart's graphic character, by noting the shapes of its outlines and the flow representation itself, the user can scan even a lengthy flowchart and rapidly grasp the dynamics of the proposed program.

Using the structural detail of the design document and the flowchart, the specialist information provided by the subject-matter experts can be organized into a script outline. The dialogue and action of the script is written around this framework.

INSTRUCTIONAL DESIGN

There are people who would not think of executing an interactive program without an instructional designer's input, and there are others who deny their necessity. Some successful disc producers do not know what instructional design is; and others claim they provide, at best, distractions from the producer's task. The truth is: none of the above, and all of the above. There is no need to feel paralyzed in the absence of a Ph.D. in instructional design. However, instructional design concerns itself with matters of sequencing, testing and remediation that no trainer, and possibly no marketer, dares ignore. It may not be necessary to hire a $400-a-day instructional design specialist for your project, but someone who fully understands the users' needs must take on the role of instructional designer if you want the cognitive structure of the interactive product to be acceptable.

A producer does not have to be intimidated by the prodigious amount of paperwork that is usually produced by instructional design people. Instructional people, after all, come from a world that measures output in volume of paper—producers don't. Have courage, and do not confuse mounds of words and symbols with the product. How you get there is not as important as achieving your goal; most vital is keeping the goals of your disc project clearly in mind.

I have produced more than 40 discs, half of them involving training, and I have worked with instructional design specialists on only a few of these, and then usually at client request. My technique in instructional design is to clarify the teaching objectives of the disc at the outset, with client input, and then thoroughly "acculturate" the scriptwriter into the operational characteristics of the proposed delivery system. Thereafter, the scriptwriter and my colleagues in disc production take the role of an IVD design team, optimizing use of the disc's interactive features while assuring the responsiveness of our script material to client needs. Only if there were extensive testing or a need to track user performance would I feel the need to retain an instructional design specialist.

More important to the development of your interactive videodisc may be the IVD designer. This is not to suggest that working with instructional designers is less than pleasant. Usually, instructional design people are refreshingly clear about how to sequence materials to achieve a defined objective; and they are downright gifted at clarifying those objectives. Moreover, in my experience, instructional design specialists are quick to pick up on the advantages of a visual medium to accomplish their goals. In a training disc project I had worked on for a major insurance company, the videodisc team came up with a visually attractive graphic in which four large blocks constituted a vaulted arch. The subject matter of the training disc concerned a computer system whose acronym was "ARCH." Earlier instructional design studies had revealed that six decision steps were crucial to the performance of the trainee's tasks, but the four-block arch was so attractive and compelling that the instructional sequence was re-structured to specify

"There Is No Need to Feel Paralyzed in the Absence of a Ph.D. in Instructional Design."

four primary activities in order to conform with our video design. Later, both print materials and computer-based tutorials used the four-block arch graphic as a mnemonic device and an illustrative key to successive lesson segments.

Authoring Languages and Systems

One way to create a cognitive structure for your disc (with or without instructional design people) is to use an authoring language or system.

Authoring languages are high-level computer languages, distantly removed from the binary strings of ones and zeros that computers understand. An authoring language makes it easier to write the computer portion of interactive videodisc courseware, particularly if the task involves computer-generated text frames. This is because authoring languages provide commands that easily manipulate text, branch to different sections of the program depending on user response, provide reinforcement and remedial material, and keep records of user performance. Graphic creation and animation are facilitated with authoring languages, and the commands to run an external videodisc player within these languages exist as a type of shorthand.

Authoring languages are also generally easier for the novice to learn than other computer languages because their commands are often in English and their written programs are relatively easy to read.

Authoring Systems

Authoring systems are a step above authoring languages in the hierarchy of computer languages. They often have several predetermined instructional strategies that can be selected from a menu. Other computer languages do not have these capabilities, although other languages may be better at performing calculations and manipulating large quantities of data. It should be noted that although authoring systems may be easier for the novice to use, they may limit the author's choices in developing courseware. In this environment, if you choose to do something that is not on the system's menu, it may be difficult or impossible to proceed. It is also necessary to choose a language that will run on your intended system. Select a package that is suited to your needs and to your proposed hardware.

Often, there is an appearance of savings in the use of authoring languages since they offer some clear time economies. But consider the costs of tying up your instructional design personnel or your subject-matter experts in this exercise—and compute again the "savings." Besides labor costs, there are several other crucial questions you must ask before selecting an authoring language: Can you update materials easily? Does the language generate truly individualized instructional strategies and do these elicit active interest on the part of users in the instructional environment? Does the language actually enhance author productivity? Can the language be moved from one computer system to another? And, can it be extended to accommodate new features and new facilities?

Many experts in the disc field question the value of authoring languages, citing their limitations in terms of flexibility and portability to other systems. Moreover, there is a dearth of authoring languages for Level 3 systems. To many experts, the more elegant and cost-effective solution is to add a computer expert to the creative team and let that person author a custom control system for the given application. Someday—maybe—there will be enough tasks, and maybe enough generic courseware, to justify democratizing the computer control activity. That day has not dawned. Until then, a resident or accessible computer expert is probably your best recourse.

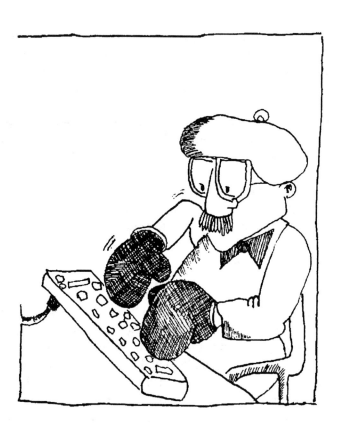

"Many Experts in the Disc Field Question the Value of Authoring Languages, Citing Their Limitations In Terms of Flexibility. They Have Been Termed 'Great Big Furry Gloves,' Not the Ultimate In Fine Control."

THE IVD DESIGNER

The essence of interactive video is the invisibility of the product. Even at the end of the producer's work—1-inch pre-master in hand—the final shape of the program will probably in no way resemble a straightforward playing of the tape. The user becomes, in a sense, the final editor, and the resulting permutations of material contained in the program may never have been seen before.

Somehow, the IVD designer must foresee every eventuality, or at least comprehend the probable outcome of each and every combination of running sequences, stills and computer-generated text/graphic overlays. One linear sequence may switch to another under certain conditions: Are both shot in close-up? Will they "cut" (i.e., look right) when and if juxtaposed? Will a sequence end in a video-recorded text frame over which the controlling computer can add its own text—now unreadable in a welter of colliding print styles? If no one manages the full range of these possibilities, the unforeseen will be seen, and embarrassment or system failure can result.

The IVD designer's role is to optimize and manage the interactive program's performance, setting forth the overall flowchart for the disc and its accompanying computer program (if Level 3). The IVD designer must visualize where every branch will lead and how every sequence will interconnect with any possible still or linear segment to follow. This modeling process can be aided by storyboards, which assist in the visualization of chains of linear sequences and/or stills. At present, there are only a few tools beyond charts and file cards to aid the designer in visualizing possible production outcomes.

When the tape has been produced, there are systems (such as the Sony Emulator) which can help IVD producers foresee the nonlinear performance of their discs. But in the pre-production stage, simulation and modeling tools are limited.

In practice, producers are usually limited to: tacking up file cards (each representing a single still or linear sequence); thumbing through loose-leaf notebooks with control pages dedicated to individual disc elements (such as stills, overlaid text frames, or motion segments); using "idea processor" software such as the *MORE* program for the Macintosh PC to see detailed outlines and flow diagrams of an interactive program's progress; or using *HyperCard* to relate the many elements in an IVD production database. MacroMind's *Director* software package offers, at long last, all of these capabilities, and a good bit more, in one place. And, new products (e.g., a powerful interactive multimedia production control program currently named *ScreenPlay*) will soon crowd this long-vacant space.

HyperCard

HyperCard, running on a Macintosh computer, permits a designer to work with electronic file cards. Ideas or script elements can be described, in words or pictures, on individual screen "cards." These cards can be organized into "stacks" which can be linked and cross-referenced. *HyperCard* is probably the best commonly available design aid for IVD producers. It comes with a video toolkit containing the drivers for nearly all videodisc players.

Templates

Other tools and techniques exist—touch screen systems, computer authoring packages and bar-code reading disc control systems—that have fixed menus of interactivity built into them. These have certain speed-of-production and ease-of-operation advantages. Since there are only finite ways in which a problem can be presented, and there are only limited results of a user action, it is relatively easy for a disc designer to lay out instructional or expository sequences, and the tests or activities that follow, using design templates. Be aware, however, that there are some real problems attached to this approach to IVD design.

Action Code

This is not to say that there haven't been success stories with templates, such as Action Code, a product of the National Education and Training Corp. The idea behind Action Code was to create a computer-controlled disc training system without a keyboard. The target audience was blue-collar and had a strong aversion to PC-type keyboards. How, then, to permit interaction? There are affordable touch screens, so one was integrated into the system.

The National Education and Training Corp. is a publisher of educational textbooks. An important goal in their corporate development was to tie videodisc to textbook-delivered training. The trick (developed by Perceptronics, one-time leaders in the production of disc-based simulators) to accomplish that end was to print bar codes in the textbooks. The bar codes, which could be scanned by the trainees, allowed a variety of inputs including answers to test questions to be presented to the system.

The Action Code system permitted a limited range of outcomes, and it had a limited assortment of design options. These options—such as branching to a short motion segment to provide remedial information following an incorrect answer—were termed "templates." And like all templates (or conceptual cookie cutters) they offered the advantage of speed at a cost in flexibility.

A program writer simply needed to designate the template for a given response, for example, four touch-screen points in a vertical line, with only one correct point. The rest of the work entailed in coding the control program was mechanical in nature— generation of appropriate bar codes—but not all the world of instructional possibilities could dance through the eye of this needle. In fact, only a very limited range of options fit the templates.

So, although the National Education and Training Corporation remains the biggest IVD-based training enterprise, servicing a broad range of industrial customers, it offers some lessons in what not to do if you want flexibility in the design of training videodiscs. Templates in any form—and this includes most authoring systems—are necessarily exclusive of most options. Be wary in your selection of shorthands. The flexibility you lose will surely be your own.

There is a parallel disadvantage to using templates, from the user side. It is highly limited, usually, in the range of available outcomes, and it becomes predictable and boring rather fast. Often, a template will deliver the hapless user to a painfully repetitive audio or still sequence that says, "No. That's wrong. Try again." This is what David Hon has termed "death in the still frame."

Sometimes it only makes sense to take advantage of an authoring system that is ideally suited to the application environment. But don't be lazy. All your production efforts will be for naught if the product is a pallid and dimensionless thing clearly better suited to getting your programmer home to dinner than to delighting and enlightening the user.

THE PRODUCTION BOOK

Whatever the authoring tools and templates mobilized, the best way for a producer to keep track of what is going on within his or her production is to keep a production book. (See Figure 5.3.) It is another important aspect of documentation. The production book, like the design document from which it grows, is a loose-leaf notebook with a single page dedicated to each motion scene or still frame on the disc. The primary use of this document is to make sure that everything gets onto the pre-master in the proper order, and for the correct duration.

I have seen some folks make a fetish of the book, filling it with every tidbit of information imaginable—including the various routes that have brought the user to a particular sequence, the learning objective that the sequence addresses and the branches that may follow. This information no doubt has value, but is such a clutter of raw data that Einstein would be overwhelmed by it—let alone some producer who is wildly thumbing through his or her book to find out if they forgot to shoot a scene. Here, in truth, less is more.

I believe that a production book should be a linear, paper model of the layout of the disc. It starts with the image that belongs at time code 01:00:00:01 and goes through the image that is to be placed at time code 01:30:00:00. The book should have provision for drawing little thumbnail storyboard sketches that describe each image, whether it is a still or motion sequence. It should cite the time code start and stop points (is it one frame long or 14 minutes and 20 seconds?). It should note any action points that must be encoded on the disc: Is this a chapter stop? Or, if a person makes a choice at this point, where are they to be branched?

I think it unwise to try to cram into this video production tool all of the architecture of the computer program that may later run the disc. Recall that from the video producer's point of view, a Level 3 disc and a Level 1 disc are nearly indistinguishable. It may be counter-productive to burden your producer, or yourself, with all the complex details of how the operating program will later work, when at this moment of disc production one should simply focus on getting the right images in the right order on the disc.

"Tiny steps for tiny feet," a brilliant colleague used to chide. It's humbling advice,

Figure 5.3: Sample Production Book Page

Event Number ☐   Communications Objective: _____

Still Frame ☐   Motion Sequence ☐

Transition	In	Out	Chapter/Module:	Screen:	Page:
Cut					
Wipe					
Dissolve					
Efx					
Mosaic					

Input SMPTE Address:

IN: ___ : ___ : ___ : ___

OUT: ___ : ___ : ___ : ___

Pre-Master SMPTE Address:

IN: ___ : ___ : ___ : ___

OUT: ___ : ___ : ___ : ___

Storyboard Illustration of Frame/Sequence

Video ☐   Audio ☐

Starting   No. of Frames:

Channel 1:   Channel 2:

Wait ☐   Continuous ☐   Hold ☐

Animated Objects

Video: _____

_____

Channel 1: _____

_____

Channel 2: _____

_____

CONDITIONS

If _____ Then _____

If _____ Then _____

Special Instructions/Programmer's Notes _____

_____

Branch to: _____

Access Mode
Bar Code ☐   Key ☐

Timed: _____ Sec.

and it's worth internalizing for this medium.

The bottom line about documentation: it's a coordination tool and a device for command and control. You need only enough to do the job right. An encyclopedic production book that proves useless on the set is no accomplishment at all. The most important goal is to keep all members of the disc production team up-to-date as changes are made in dialogue, still frame content and count, and the overall operation of the disc. People must be updated concerning changes. The best way to keep track of pages in the book, or locations of scenes, is to religiously use SMPTE time code to track everything. Two things cannot go in the same place. If people are attentive to still and motion scene locations, problems in use of the disc real estate will be detected. Keep this critical location information on top and pay painstaking attention to it.

Let the computer programming team (if you're dealing with Level 3) use a parallel set of documents, keyed to your production book. Remember, at the start of the project you had a flowchart—everything keys back to that. The flowchart schematically captured how the program flows under user control. Programmers and producers are following that same high-level outline. The first page of both your production book and the programmer's manual should be the original project flowchart, updated as needed (with copies shared by all critical personnel).

A word about document coordination. The video production team may well have a script supervisor who keeps meticulous track of all changes in wording and scene execution details as they are experienced on the set. But who will coordinate the activities of the different working groups—still frame producer, computer programmers, graphics designers, and possibly print instruction manual writers—whose work must later interface with your own? The answer is: the project manager. This person, or their designee, must make certain that everyone receives the correct paperwork (read that "changes in paperwork") on time, and in a coordinated fashion. If the programmers are ignored, and so do not learn of timing changes in the linear video material, they may waste extensive effort writing codes that will have to be modified to accommodate another reality. That sort of waste is unforgivable.

To my own injunction to never be sorry, I now add Eggert's Dictum: "Don't look stupid."

CONCLUSION: FOLLOW THE DESIGN PATH

To recapitulate, the design path for videodiscs discussed in this chapter is simplified below:

- A needs analysis details the behavioral or instructional goals of the project. The task to be addressed is analyzed and specific behavioral objectives are stated.
- A technically and financially appropriate interactive delivery system is specified.

- In consultation with client subject matter experts (SMEs) and IVD design consultants (possibly including instructional designers), the overall design of the product is laid out. A flowchart of the interactive product is prepared.
- The design document is produced, detailing content, system operation and user inputs.
- Scriptwriting proceeds, based upon SME-provided information, IVD design and instructional goals.
- A production book is produced, integrating interactive designs, script and program flowchart.

6 Pre-Production and Scriptwriting

Pre-production, seen with a little perspective, is the heart and soul of production itself. As we enter this phase, task analysis has been performed, learning and behavioral objectives have been finalized, subject matter experts have been hired and the design document has been completed. Now, the script is written, the production plan developed and the crew assembled. (See Figure 6.1.)

The scheduling of these events, as well as post-production, programming, testing, packaging and delivery is also accomplished during pre-production. The producer can use time lines (see Figure 6.2) to set up a project timetable, which helps to keep a production within its budget limits and organizes retained staff (see Figure 6.3). (For further discussion of scheduling as it relates to budget considerations, see Chapter 4.)

SCRIPTWRITING

The master plan for the linear portion of the disc program is the script. Here one finds all the words to be said, all the images to be recorded, and the order in which they are to be presented. Multiple sound tracks, special effects and stop codes for video are thoroughly described in the script.

Scripts should be in all key personnel's hands, prior to shooting. It seems silly to say, but this needs to be stressed: the script should be finalized before the start of production. Invariably, in my experience, there are changes—large and small—made during the shoot. I've had one experience in which script changes were made after the shoot! (These were realized in voice-over narration.)

Figure 6.1: Design Tasks Lead Into Pre-Production

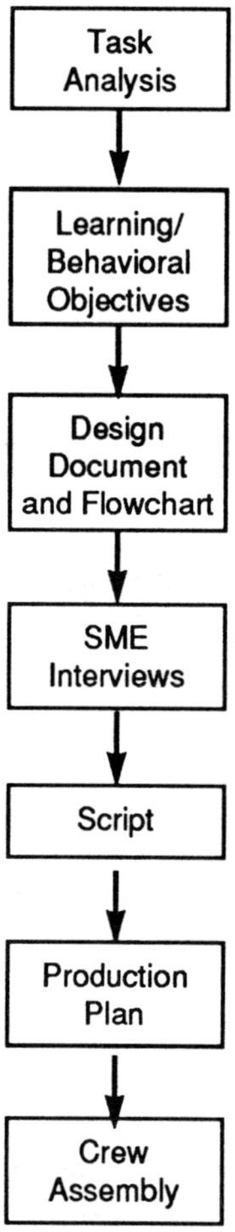

```
        ┌─────────────┐
        │    Task     │
        │  Analysis   │
        └─────────────┘
               │
               ▼
        ┌─────────────┐
        │  Learning/  │
        │  Behavioral │
        │  Objectives │
        └─────────────┘
               │
               ▼
        ┌─────────────┐
        │   Design    │
        │  Document   │
        │and Flowchart│
        └─────────────┘
               │
               ▼
        ┌─────────────┐
        │    SME      │
        │ Interviews  │
        └─────────────┘
               │
               ▼
        ┌─────────────┐
        │   Script    │
        └─────────────┘
               │
               ▼
        ┌─────────────┐
        │ Production  │
        │    Plan     │
        └─────────────┘
               │
               ▼
        ┌─────────────┐
        │    Crew     │
        │  Assembly   │
        └─────────────┘
```

The scriptwriter, a subject matter expert and/or the client's representative should be present on the set to approve script changes in real-time. The worst of all possible worlds is to be compelled to reshoot, or to shoot additional materials, after the set is struck, the crew disbanded and the budget spent.

The Scriptwriter

Disc content is provided by a scriptwriter or scripter. Working with the producer and director, the scripter first produces a short treatment similar to those written for television programs or film scripts. Settings are defined, characters sketched, and action outlined; each section or chapter is presented in a paragraph or two.

A good scriptwriter need not be a content expert. He or she can work with the subject matter experts (SMEs), discussed in Chapter 5. SMEs should be provided by the client and interviewed directly (preferably live and in person) by the scripting personnel. An informational rapport can be created, and the SMEs—usually highly flattered to be

Figure 6.2: Time Line for Large-Scale Level 3 Training Project

TIME (MONTHS)	1	2	3	4	5	6	7
PHASES							
Proposal	X X x x						
Design		x x X X x x					
Scripting			X X X X X X x x				
Pre-Production				x X X			
Production					X X X x		
Post-Production						x X X x	
Programming						x x x x X X X X	
Testing							x x X X

Note: Each X represents a week. A large X represents full-time effort; a smaller x represents part-time efforts.

Figure 6.3: Project Timetable

Phase	Milestone (ouptut)	Aug.	Sept.	Oct.	Nov.	Dec.	Jan.	Feb.
Needs analysis Course development		▮	▮					
Instructional and interactive design	Design document flowchart	▮						
Scripting	1st Draft			▮				
Script review				▮				
Rewrite	Final script				▮			
Script approval					▮			
Pre-production	Cast/crew/sets wardrobe				▮			
Video production					▮	▮		
Off-line/review, revisions	Rough edit					▮		
On-line review, revisions	Fine edit with music/effects					▮		
Computer-based course authoring programming	Screen design and content			▮	▮	▮	▮	
Integration of videodisc	Total program 1st draft						▮	
Total program test and debug	Finished program							▮
Replication and packaging								▮
Delivery								▮

involved in what we call show biz—become an ongoing link throughout the production. (SMEs should be reachable by phone for script additions and revisions. They should also be available to read the final script and call attention to any glaring informational errors).

Gathering Information from the Field

After the SMEs have offered their information, it is useful—vital, in fact—for scriptwriters to visit the subjects of their creative efforts. There is no substitute for a journey into the field, to see the people, places, tasks and products of those who will be trained, sold to, or entertained via IVD. In an instructional project, for example, there is a texture and a tone that is specific to a place (and its inhabitants) that can never be conveyed save by actually standing in the space. The best training arises, in part, from resemblance to actualities to be encountered. No one is well served by a slick production that bears little resemblance to its real subject matter.

On one of my own disc projects, a training program for a midwestern insurance company, the producer, director and scriptwriter spent three days on site interviewing SMEs—taking marginal notes all the while on how they dressed, how their office space was decorated, and what slang terms they used to describe company materials which would be used daily by the trainees. All of these elements were integrated into the live-action visuals of the training disc. In fact, actual words of advice from current workers were integrated into the training disc in documentary-like "testimonial" segments. These were placed at the start or end of a training section and generated tremendous impact and appeal.

To grasp the visual feel of your subject matter, to mirror its costumes, personalities, ambience, lighting, and even its purpose—go to the real-world referent—visit the site, meet the inhabitants. Besides being a competent interactive product creator, become a video anthropologist. Your client will thank you.

The Script Begins to Take Shape

After encountering and debriefing the SMEs, the scripters should retreat to review their testimony and the documents that elaborate it. This information will become the content of the script. The words may appear as dialogue or voice-over narration, as text on the screen or annotations on charts. However the data will be presented in its final form, at this point it has been articulated. The next step is to analyze and edit it.

There are usually two parts to every script: the video script and the audio script. The video script may contain stills, motion pictorial sequences or text. The audio script contains material to accompany these segments.

It is prudent to divide your script pages into at least three parts: left-third, entitled "Video"; middle-third, entitled "Audio Channel 1"; and right-third, entitled "Audio

Channel 2." This division will help remind the scripter and producer of the powerful efficiency offered by use of multiple sound tracks, which may, of course, contain the same information but possibly very different data. Atop each script page, it is wise to number the linear sequence or still, noting its position (in time code numbers) on the disc, and also pointing out where the viewer has just come from. I like to add a small box for thumbnail storyboard sketches (stick figures are fine here), to force myself, and my cohorts, to visualize how scenes will work when they are juxtaposed under user control. (See Figure 6.4.)

PRE-PRODUCTION NUTS AND BOLTS

The IVD design has been approved by the client and a script is nearing completion (final revisions and approvals are all that remain). At this point, we choose the director, determine the cast, pick the location and hire the crew. We secure the equipment and prepare a detailed breakdown of the script—so detailed that we may never be more than an hour off our projected shooting schedule. In pre-production, we translate our design and our script into human form. We see the product, though it is not yet on tape—it moves before us at casting call-backs, reading our words aloud; it takes shape in dim rooms that will soon be bright and a bit too busy (the sets), as art directors measure the length of walls to be built and lighting directors check power tie-ins.

Pre-production is, or ought to be, every activity that underlies, shapes and readies the production up through the load-in of sets and equipment to the location. Pre-pro must cover the details of logistical support for your production, along with technical and creative support. Cars must be engaged, hotel rooms must be reserved, and catering has to be arranged.

Some words about catering: if the call is early, you should provide a breakfast— rolls, coffee, tea, fruit, and some protein additions such as cheese and hard-boiled eggs. If you haven't planned to release your crew for lunch, which can cost you a "rope start" even if everyone adheres strictly to a one-hour break schedule, you have to provide lunch. And if you go late, a hot dinner is considered humane, if not obligatory. Finally, the words "wrap" and "beer" are often paired in the production biz. A couple of cases of goodly brew are *de rigueur* when shooting is completed. These elements of a production are usually as vital as any technical service. They require consideration in advance, to save money and conserve management attention for the inevitable crises and creative challenges of production.

CREATING A CREW AND PRODUCTION TEAM

Let us now turn our discussion to the hiring sequence during the pre-production phase that results in formation of a crew and production team. (See Figures 6.5, 6.6 and 6.7.)

Figure 6.4: Two-Column Script Page Created by David McCall for Murder Anyone? Mystery Disc I
(Original text written by Hy Conrad)

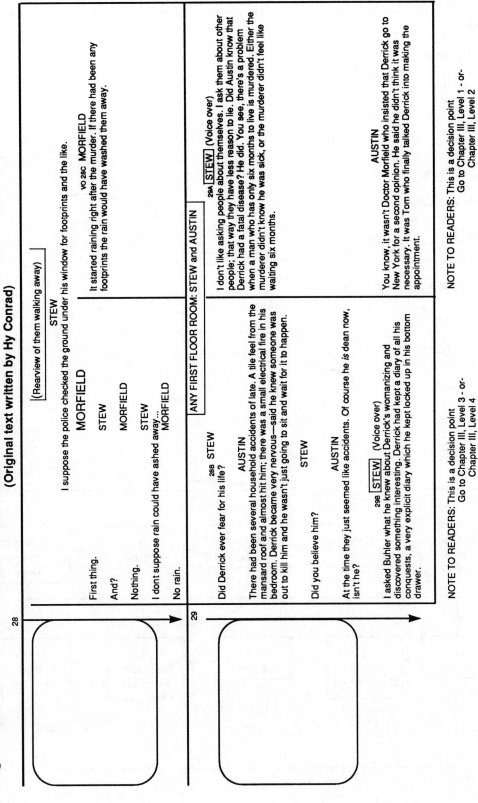

28

[Rearview of them walking away]

STEW

I suppose the police checked the ground under his window for footprints and the like.

MORFIELD

First thing.

STEW

And?

MORFIELD

Nothing.

STEW

I dont suppose rain could have ashed away...

MORFIELD

No rain.

vo 28c MORFIELD

It started raining right after the murder. If there had been any footprints the rain would have washed them away.

29

ANY FIRST FLOOR ROOM: STEW and AUSTIN

28B STEW

Did Derrick ever fear for his life?

AUSTIN

There had been several household accidents of late. A tile feel from the mansard roof and almost hit him; there was a small electrical fire in his bedroom. Derrick became very nervous—said he knew someone was out to kill him and he wasn't just going to sit and wait for it to happen.

STEW

Did you believe him?

AUSTIN

At the time they just seemed like accidents. Of course he *is* dean now, isn't he?

29B STEW (Voice over)

I asked Buhler what he knew about Derrick's womanizing and discovered something interesting. Derrick had kept a diary of all his conquests, a very explicit diary which he kept locked up in his bottom drawer.

29A STEW (Voice over)

I don't like asking people about themselves. I ask them about other people; that way they have less reason to lie. Did Austin know that Derrick had a fatal disease? He did. You see, there's a problem when a man who has only six months to live is murdered. Either the murderer didn't know he was sick, or the murderer didn't feel like waiting six months.

AUSTIN

You know, it wasn't Doctor Morfield who insisted that Derrick go to New York for a second opinion. He said he didn't think it was necessary. It was Tom who finally talked Derrick into making the appointment.

NOTE TO READERS: This is a decision point
　Go to Chapter III, Level 3 - or-
　　Chapter III, Level 4

NOTE TO READERS: This is a decision point
　Go to Chapter III, Level 1 - or-
　　Chapter III, Level 2

Figure 6.5: Pre-Production Personnel

Writer

IVD Designer

Programmer

Client Representative

Producer

Project Manager

Production Manager

Director

Still Frame Producer

Production Designer

Art Director

Post-Production Supervisor

Pre-Pro Activities

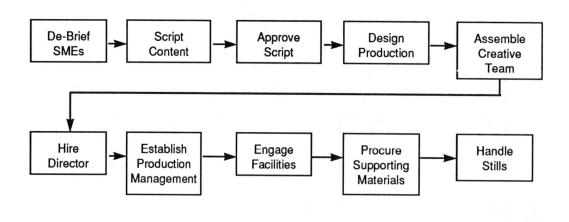

Figure 6.6: Pre-Production Personnel - Flow of Information - Sequence of Personnel Into Project

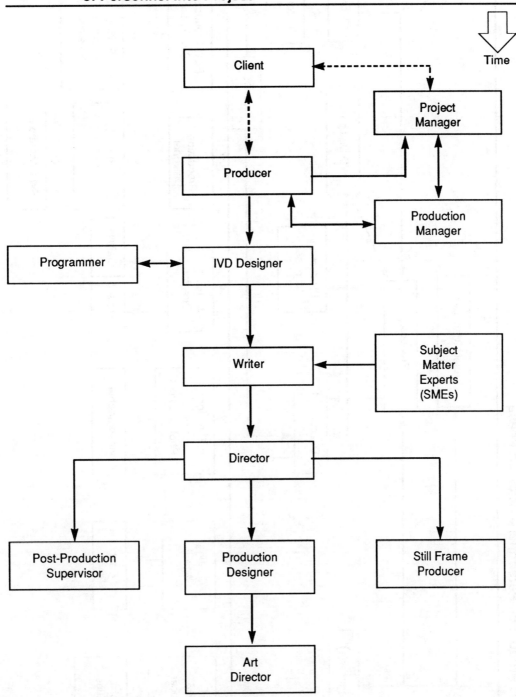

Figure 6.7: Production Personnel — Kick-off Phase

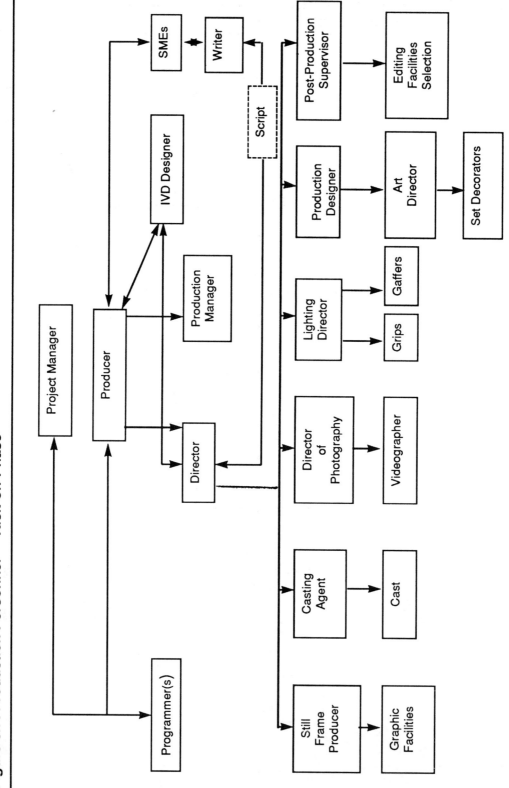

The Project Manager and Producer

First in, of course, are the project manager and producer. (They may be the same person.) The producer, once oriented, should retain a production manager. The production manager is the keeper of the purse—the one person who can (and must) say "no" to everyone. At the very start, the production manager and producer should sit down and come to a detailed understanding of creative goals and financial limits. The creative vision will at once begin to be tempered by fiscal realities, and the "green eyeshade" needs to be recruited to the expressive and interactive goals of the project. Once the realizable vision of the product has been agreed upon, its implementation becomes the focus.

Traditionally, the key role of the producer was to hire the director. Indeed, in the great organization chart in the sky, the sole sacred role of the producer was the selection of the director. I believe that the producer in interactive production must be a more proactive soul. In interactive, the producer may also act as project manager, with all the attendant burdens. In the creative domain, it has been said that film is the director's medium, while television is the producer's—I believe this to be doubly so for interactive applications of video. The producer's vision, style and imprint are everywhere. Video's odd ability to X-ray the intentions and spirit of those behind the camera amplifies the presence of the producer, in silent but potent ways, and imprints the product with his or her ineffable qualities.

The Director

Nonetheless, the producer hires the director, and thereby hangs a decisive part of the tale. The director runs the show. He or she will decide upon the cast and will usually have a voice in all department head selections. When shooting is underway, in order to keep communications straight and accountability clear, all changes, suggestions and client input are routed through the director for implementation on the set.

The director's key pre-pro moves, besides internalizing the script, are to hire a production designer and/or art director, and to initiate casting.

The Production Designer

The production designer conceives the overall "look" of the piece and moves to achieve this through selection of location, set design and decoration, and specification of props.

The Call for Actors

The casting agent, hired by the director, puts out a "call" for actors who fit the script's requirements. These actors may submit "head-sheets"—facial photos with resumes attached—for review by the director. Plausible candidates are summoned to a "reading," at which the majority are eliminated, and a few are selected for "call-backs." At the call-back, where further extemporaneous reading of the script occurs, the final cast members are selected. Ordinarily, this screening is attended by the director, the producer and usually a client representative.

The Still-Frame Producer

A decision must be made about the volume of graphics and stills to be included in the product. If producer, director and project manager concur, a special still-frame producer may be selected to manage that phase of the work. This person is responsible for the timely, accurate and aesthetically apt execution of all graphic and text frames that are to be part of the disc. The still-frame producer retains graphic artists to prepare the work, and investigates competent and cost-competitive facilities to produce the required graphics and text. The costs and turn-around times of these facilities are reported to the production manager for review by the producer.

Selecting Production and Post-Production Facilities

The production manager must ascertain, with input from the producer and director, which facilities are to be used for production and post-production. Detailed time requirement assessments for each phase of work must be plotted against hourly and daily rates. A facilities cost for the job must be crystallized. (Usually, this is an estimate, but one with a fair amount of detail and a high degree of predictive value).

The Computer Programmers

Finally, the computer programmers should set upon their work. Aided by the IVD design documents and flowcharts, the programmers flesh out their designs, noting any glitches in logic structure, real-estate or timing requirements that may emerge. These are reported to the project manager and the producer.

Writers and subject matter experts hover at the borders of all this activity, standing by to offer advice, alternatives or additional input, as circumstances arise.

This is the loom of disc pre-production.

CONCLUSION

At the end of the pre-production phase, a production has been planned, crystallized around your original design (encoded in a flowchart). To keep your documentation up-to-date, the production book now includes a multi-column script, a script breakdown, a shot list, cast list, crew call sheet and all necessary contact information to reach each person on the team. Copies of the book are held by the producer, the director and the project manager. Now, an even greater adventure beckons—production.

7 Hands-On Production for Discs

"Knowledge is knowing that we cannot know." —Ralph Waldo Emerson

The script is approved, the budget is in hand, and the production plan is in focus. How do we make this thing? What must we look out for? What is different from "normal" linear production?

At this point, you should have in hand a fairly detailed time map of the layout of the disc. This real-estate "map" dictates how long the linear sequences should be, how many still frames are required, and how linear and still materials will mesh. These requirements are to be treated as sacrosanct. They should be written on the production book's pages and on the scripts. They should be hammered into the director's head and emblazoned on the assistant director's and production manager's minds. For lo, if they get lost, scotch tape won't help, and truly profound apologies will be in order.

PAY ATTENTION TO DETAIL

When *Murder, Anyone?* was produced, it was commonly believed that the consumer laser disc could not be relied upon to offer viewable images on its outermost edge (i.e., beyond 28 minutes of material per side). So agonizing efforts were made to shoehorn 20 different motion sequences and several thousand still frames into 27 minutes and 40 seconds of running space.

By the time MysteryDisc II was produced, a certain confidence had crept into IVD circles that one could go to the edge, literally, using all 30 minutes of space. As a result, script supervisor, director and even editor relaxed, more than a little, the harsh rigor of the scene timing surveillance that characterized MysteryDisc I. The result: the first cut of *Many Roads to Murder*, not including still frames (which ran in excess of two minutes)

71

timed out at 34 minutes. The final product, after cruel cuts, ran 29 minutes and 40 seconds and, yes, images broke up on many players during the last minute of material.

McLuhan wrote "technology is explicitness." Anyone who has produced video knows that if you wink at a detail, you may find your posterior relocated. (Those batteries someone forgot to charge; the adaptor plate you neglected to obtain that allows your O'Connor head to fit the Elemack dolly you rented; etc., etc.) True in linear production, trebly so here in interactive dimensions. Everything must be faced, addressed and dealt with. There are no "details" in disc production, just fatal flaws hoping to be overlooked. But don't get nervous, being uptight may only cause you to blow the most vital item of all—smooth, open, harmonious teamwork in which everyone shares "the mission." The producer can't do this stuff alone, so you must be a gracious paranoid. Recruit everyone to your understanding that this is risky business. Tell them that loose lists sink discs. Everybody in the cast and crew should be on the *detail* detail. There's no room for mistakes here—you probably won't be able to fix them in post.

Delegate the Financial Responsibility

Find a production manager who will handle everything financial on the set, and make certain that documentation is precise. (See Figures 7.1 through 7.4.) Set up an iron-clad petty cash disbursement system. It is with proper petty cash vouchers that the crew will ransom their paychecks. A good production manager should know how to do this, without making people nervous. Get your head clear of the money, you have bigger things to worry about.

THE PRODUCTION STAFF

In the preceding chapters we discussed the responsibilities of staff members during the various phases of a disc project. The following list defines those people who are important during the production process, particularly while material is being videotaped on set.

- Producer: Oversees the entire process.
- Director: He or she knows the layout; they're disc- and interact-oriented; they're sleeping with the script on their pillow.
- Production designer: Conceptualizes the "look" of the product.
- Art director: Executes the production designer's and director's vision; has hired prop assistants and set dressers who are running about securing whatever it will take to make the set look right.
- Videographer/cameraperson: The director and videographer review the script scene-by-scene to decide on shooting angles, set-ups and lighting; also arranges for camera and supporting equipment, in consultation with the production manager.
- Lighting director: May design the lighting or may simply execute the production designer's and director's instructions; specifies lighting instruments, supporting

Figure 7.1: Sample Work Sheet for Production Budget

Development

    Script   _____
    Storyboard   _____

    Producer   _____

Pre-Production

    Production Management   _____
    Director   _____

Production

    Talent: Casting   _____
                            Audition space   _____
                            Hair and Makeup   _____

    Director   _____

    Assistant Director   _____

    Second Assistant Director   _____

    Art   Art Director   _____

                            Props   _____
                            Hardware   _____
                            Labor   _____
                            Trucking   _____
                            Consumables   _____

    Costumes   _____

    Sound   _____

    Video: Studio Rental   _____
                            Location fees   _____
                            Equipment rental:
                              Camera   _____
                              Recorder   _____
                              __days x __ (test)   _____
                              __days x __ (shoot)   _____
                            Camera mount   _____
                            Lenses   _____
                            Monitoring equipment
                              Video   _____
                              Signal   _____
                            Other (list) _____   _____
                                   _____   _____

Figure 7.1: Sample Work Sheet for Production Budget (Cont'd.)

Crew   _____
                                                                 _____
                                                                 _____
                                                                 _____

Director of Photography   _____

General Expenses   _____

Lighting Director   _____
                                  Grips   _____
                                  Gaffers   _____

Lighting Instruments   _____
Casting and Location   _____

Production Manager   _____
Production Assistants   _____
                                                                 _____

Videotape   _____

Petty Cash   _____

Post-Production

    Offline Editing   _____
    Online Editing   _____

Other

                              _____   _____
                              _____   _____
                              _____   _____

                                                    Total: _____

Figure 7.2: Budget for an Early Interactive Movie IVD Arcade Game (1983)

<u>Linear Production</u>

Three day Shoot:

Writer	$3,500
Director	4,500
Asst. Director	1,250
Lighting Director	2,500
Grips	1,600
Gaffers	1,200
Script Supervisor	1,250
Costumer	1,000
Wardrobe Asst.	500
Art Director	2,500
Set Decorator	1,000
Carpenter	750
Sound	1,500
Production Assts.	1,000
Hair	750
Asst.	300
Casting	750
Camera Operator	1,200
Asst. Camera Operator	800
Crew Total:	**$27,850**

Cast	6,750
Costumes	2,500
Location fee	6,000
Set dressings	4,000
Consumables	2,500
Extras	1,125
Contingency (10%)	5,000
Lighting rental	2,000
Dolly and grip	1,750
Camera and VCR	3,000
Production Manager	1,500
Travel	1,500

<u>IVD Elements:</u>

<u>Level 3 Programming</u>

Programmers	$5,000

<u>Archival Footage</u>

Licensing	15,000
Research and dub	5,000

<u>Product Integration</u>

Software testing	3,500
Enclosure proto.	4,000
Travel to demo	2,000

<u>Post-Production</u>

Offline edit	3,500
Online edit	6,500

<u>Pressing</u>

Fast turn-around (aluminum backed)	3,000

Project Total:	**$112,975**

Figure 7.3: Fortune 50 Corporate Level 3 4-Disc Training Package

Contractual Obligations: Script authoring
Image creation—motion and still
Disc layout and pre-mastering

Pre-production Services:

Design consultation
Script conferences
Scriptwriting
Production Planning

Production:
Above-the-line personnel
Project director
Director
Producer
Writers $24,000

Linear Motion Production:
Production facilities and crew   68,000
Art department   15,000
Cast   12,700
Transportation   1,400
Tape stock   1,300
Wardrobe   1,500
Food   1,800
Prod. office supplies   650
Misc.—furniture rental consumables   1,700
Shipping   600
Client entertainment   500
                                                    _____
                                                     $129,150

Still Frames Production:
(1000 text frames)

Graphic layout and design $2,500
Character generator—input   2,500
Character generator—capture on videotape   3,500

Post-Production
Offline (linear segments $5,000   (3 weeks)
Online
1. Sub-master assembly   12,000   (6 days)
2. Text frame integration   5,500   (3 days)
3. Pre-master assembly including graphics   6,000   (3 days)

                                          Total: $166,150

Figure 7.4: Some Estimated Costs

Production Facilities and Support

Studio or Location

      Trucking

      Craft services (food)

      Cars

Equipment	Per Day
Cameras	$650 to $1,000
Tripod; dolly	$ 50 to $100
Head	$ 60 to $100
Batteries	$ 35 each
AC power	$ 50
Monitors	$100 each
Videotape recorders	$150 to $500
Power supplies	$ 75
Tape	$ 60/ hour
ATR and audio mixer	$150 to $250
Mics; headphones; cables; intercom	$ 50 to $200
Lights	$350 to $750
Grip equipment	$150 to $250
Lamps; gels; diffusers; etc.	$150
Petty cash	$250 to $500

equipment and power tie-in requirements; orders these through the production manager.

- Gaffers: Also known as the "second electrics," the gaffers work for the lighting director, hanging and trimming lights, arranging power lines, etc.
- Grips: These staff members handle heavy equipment on the set, assisting both the camera and lighting departments; they move dollies, raise, lower and relocate lighting stands, etc.
- Sound: Can be a one or two person job, often the latter. There is a recordist who watches levels on a VU meter, and a "boom" who directs the microphone(s) toward the person(s) on camera.
- Assistant director (AD): "Runs the set," as it is said; this person is the timekeeper who prods the director, speaks to each department head, and summons the talent—while gently but firmly keeping the shoot on schedule.
- Script supervisor: Works closely with the assistant director and director, watching every scene, making certain that every line is clearly captured, that every new camera angle is cutable, that props don't move in the background and costumes do not get rearranged between shots. The script supervisor is the keeper of continuity and the guarantor that every detail is covered, and every line recorded properly—a vital role in disc production.
- Costumer: Possibly assisted by dressers if the cast is large; the costumer designs, secures and prepares all clothing worn on camera. This function is separate from the art department, but clearly closely related.
- Hair and makeup: Not to be overlooked.
- Production assistants: These staff members do anything and everything, since they support every department. Key personnel can stay on the set while these "tele-peons" support the work on the outside. Production assistants run to pick up more tape, forgotten props, replacement equipment, additional gels, lunch, wrap beer—you name it, you'll need it and you won't want to send your assistant director out to get it. They also work their butts off on the set.
- Talent: Waits around; and waits some more; and then they're called; and then they wait again. It's very glamorous.
- Tape operator: This person, in the age of Betacam recorders, can be useful in the preparation of shot logs on the set, noting take numbers, and time-code start and stop points.
- Still-frame coordinator: May be on the job elsewhere (not on the set).
- Post-production coordinator: Can be a real asset on the set. If you need to know whether or not you have enough wide shot cutaways to cover the second sound track, this is the person who will live with the answer. If he or she is not there, you won't be able to answer these questions while there is an opportunity to rectify the problem.
- Location manager: Keeps things smooth with whomever owns the location; addresses logistical requirements; and may handle craft services (i.e., meals, etc).

Of course, the production manager and the project manager should be available during the entire production process. After all, this is it. Here is where most of the money is spent. The choices that are made during production are, or ought to be, irreversible. So keep project management and client approval personnel close at hand.

THE SHOOT

Departmental Responsibilities

The shoot activity breaks down into the responsibility of discrete departments as follows:

- Art
- Camera
- Sound
- Lighting and grip
- Talent
- Costumes
- Hair and makeup
- Location services

In managing the shoot, the director addresses department heads—art director, production designer, lighting director, etc.—when he or she requires a change or a recommendation. Conversely, any professional on the set will ordinarily direct their suggestion through their department chief. A tightly run set with 40 bodies is not a place to shout out some great idea.

Develop a Laundry List

Before the shoot, it is important to go over the list of facilities that will be required. This is done to determine what you have and what you will need, and can be done by checking with each department. Or, you can simply create a laundry list. The following is a sample laundry list that has been annotated by department.

Art Department

Props and set dressings—these will vary according to project.

Camera Department

Camera "head"—the main device; the electronic package that dissects light into its brightness and color components. Care in camera selection is prudent. I recommend as high-ended a camera as the budget will allow. (Personally, in the not distant past, I only used the Ikegami HL79E.) Currently, there are several chip cameras that do the job; chip cameras cannot quite match the top end of tube devices, yet, but they're getting there, and have compensatory advantages—lighter weight, longer battery life, no sensitivity to direct sunburn, and no need to re-register on location (i.e., they can take more damage).

Lenses—motorized zoom lenses, wide-angle zoomers, and prime lenses.

Camera control unit—used in some but not all cases; allows user to "paint" the image, emphasizing particular color tonalities and altering the contrast range.

Waveform monitor—allows engineering personnel to see the profile of the electronic output of the camera; this is important in disc since brightness levels must not exceed certain thresholds.

Vectorscope—allows precise engineering control of the colorimetry of the recorded signal; usually comes in a companion package with the waveform monitor.

Display monitor—allows you to see the result of your recording efforts, but not precisely. (Engineers laughingly explain that the American NTSC color television standard really refers to "Never Twice the Same Color." Component video recording is helping to alleviate this problem.)

Batteries—depending on your hardware configuration, these nickel-cadmium devices can give you up to several hours of operation. Be redundant on this item; pack four if you think you'll need two.

AC unit—gives you the ability to run your camera and deck off wall sockets. Be cautious about the stability of your location's AC power supply, and watch out for crossing video or audio lines with your AC. There's a lot of video noise that can creep into your system on AC.

Intercom—allows the director and camera folk to speak with each other; can be vital if you're doing live switching between several cameras, or if you are performing a multiple-camera isolated recording.

Tripods—these come in several sizes and flavors. The head atop could turn out to be one of your most vital pieces of equipment. Fluid heads, capable of mounting at least 30 pounds of camera, are recommended.

Adaptor plates—allow particular heads to work on specific dollies; these are the vital link for tying camera to tripod. Make sure you have the right device; you won't want to learn otherwise with 40 people waiting around on the set.

Dollies—allow rolling shots and dramatic axial motions on "jib arms" that extend off a center post. A dolly is vital if you wish to track along with a moving spokesperson, or if you propose to roll through a plant. They are heavy pieces of equipment (which yields their stability), so they require a dolly grip (an extra person) for movement and set-up. The Steadicam and Steadicam Jr. are taking more of the dolly's role in electronic field production.

Videotape recorder (VTR)—you'll want to record on something. The options range from 90 pound 1-inch format "portable" devices to five pound Betacam or Pro Hi-8 recorders, with 20 pound 3/4-inch VTRs in the middle. Your decision will depend on available editing equipment, cost parameters, and personal preferences regarding quality and manageability. (Betacam SP is the preferred field production standard, at this time.) Because disc is a more transparent medium than tape, offering more detail and higher quality imagery, it makes sense to use the very best recording gear in original production. So-called "high-particle," metal backed 3/4-inch tape or the newer higher quality SP 1/2-inch formats are recommended for field recording. One-inch Type C, a broadcast standard, is nice, of course, but it's costly and for most tasks, it's overkill. Industrial 1/2-inch or second generation 3/4-inch will usually result in an inferior product. Hi-8 (improved 8mm) and S-VHS are bidding to become professional recording systems. At this writing, they show promise and can have real use for training, point-of-information or exhibit projects. They will soon be up to the standards of commercial replicate discs.

Sound Department

Microphones; boom and mic stands; audio mixer; Nagra or other audiotape recorder (optional)

Lighting Department

Instruments—light "heads" come in two basic varieties: halogen (HMI) or quartz lamps. HMIs run cooler, provide a bluer light (more akin to sunlight) and use less power. They are costlier to rent, but you need fewer of them to light a scene. Quartz lights are rated by wattage, and are differentiated by the degree to which they focus or flood the light emitted.

Gels—are used to color the light.

Barn doors and flags—cut and shape light, hide shadows and obscure flares (bright reflection points).

Power tie-in—attaches equipment to a location's mains. It is often safer to distribute your own power than to run lines to various wall outlets. A power tie-in will help you to avoid sudden and disastrous outages, and to sidestep line voltage variations.

Stands—of various heights and girths support fixtures.

Sandbags—help to keep the stands from falling on the cast and crew.

Clips, poles, assorted hardware—gaffers will need these to rig lighting from seemingly impossible perches.

Grip Department

Dollies; tripods; high-hat—ankle-high camera mount; stands; sandbags; more clips. (Even a small shoot requires a lot of lighting and grip gear, see Figure 7.5.)

Set Dynamics

Each of your departments will require some modicum of space on the set or location within which they can muster and marshal equipment. Certainly, cast, costumes, and hair and makeup will require some private space. Blankets will do if walls are unavailable, but separate spaces should be established. Talent needs to dress, be coiffured and cool out in peace. Allocate this space immediately when you arrive at the set. The load-in at the location may have already included set up of costume's area and a "green room" for talent. If the schedule is tight, you may have to do this the morning of the first day of shooting.

Territoriality is a fact. Honor it with thoughtful planning and be decisive in your moves. Each of your departments will need a corner to dump their gear—plan for this, and protect the allocation. Tranquility and efficiency will depend on it.

The Shoot Begins: Places Everyone!

We are loaded in and have begun to set up. (See Figure 7.6.) Nothing—nothing on Earth—is more exciting than a set being rigged by sure-footed pros moving at top speed. This is the adrenalin rush to which we who have committed decades to production are addicted.

A set is an organism. It is the closest that humans come to bee- or ant-like activity. Lives are usually not at stake, so there are no distractions from the tasks at hand. A natural and silent transfer of autonomy occurs to the designated leaders.

The producer's role is to stay on top of developments and problem-solve swiftly and surely. Do not leave the set. If the client has to meet with you, the set is where the meeting must be. (See Figure 7.7.) If something is needed, send a production assistant.

The media manager's or executive producer's job is to hover with a certain royal detachment, hoping and praying that the money is being spent well. The torch has been passed to the producer. Lines of authority must be strictly adhered to.

The best director and the best producer proceed with their work in near silence. (See Figure 7.8.) Conversation is hushed across the set. Hammers pound. Cables are stretched. Men and women hurry purposefully about, tool belts slung from waists. A once-empty space, however vast, becomes suddenly full. Centers of activity are defined. Cameras are spotted. The talent arrives and is whisked into private precincts.

Figure 7.5: Lighting Director at Work on the Set

Figure 7.6: Set-up of Office Set for a Large IVD Training Production

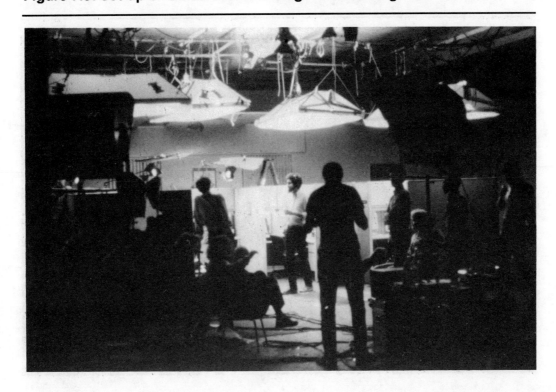

Figure 7.7: A Hurried Script Conference on the Set

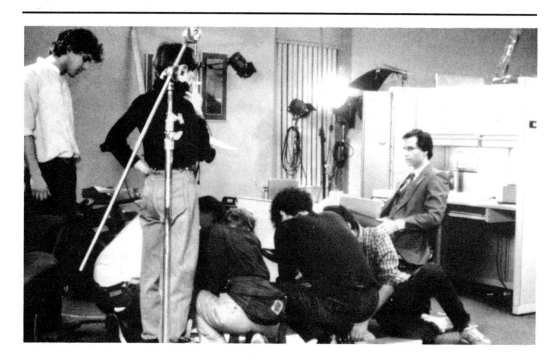

Figure 7.8: IVD Game in Production

In this purposeful welter of action, the hothead and the hysteric are anathema. You have created an organism. It has many arms and legs, but only one focus, one object, one loyalty: the product. Any ego that takes precedence to that sacred, invisible object—the product—must be shown the door.

Forty people, pulling together, can create magic—and video will capture the magic and make it apparent to the viewer. The state from which magic arises is mystical, it entails a fusion of separate beings and their talents. (This is true for computer training applications as much as for gameware or experiential simulation.)

It is up to the producer to create this magic, to mystically infuse his or her crew with a burning clarity and sense of purpose, which drives them to go for the optimal achievement of their creative goals. This can be done (only) with care, with love, with humility, and in peace.

Everyone's loyalty, first and last, must be to the product, so let us close this section with another aphorism: loose hotheads sink discs. Shoot 'em. Fast. On a set, the crew needs and wants a leader. Be one.

STILL FRAMES

The other, not so glamorous side of IVD production is the development of still frames.

It is tempting to suppose that dropping still frames onto videodisc is an easy operation. The metaphor that comes to mind is that one simply nails the required number of still frames into place and is on one's way. This is not true. Still frames are not cheap and still frames are not easy. Probably the biggest problem with placing a still frame on disc is that you will have to check it, precisely, at least four times:

1. You will have to check the still frame's proposed location to make sure it is correct.
2. You will have to lay the still frame onto the tape.
3. You will have to check to make certain that the still frame is where you put it; that it is just two fields (or however many frames) in duration; and that it is indeed in just the right place.
4. You must later return to check the frame location of all your still frames, before you release the pre-master.

There are reasons why one might return even more frequently to check the work. In a recent production, I checked each still frame location seven times. And when the disc was pressed, everything was off by one frame. Off because I had improperly noted, on paper, the first active frame of video. My advice: follow pre-master guidelines to the letter.

> The first active frame of
> video is 01:00:00:01

I speak here of individual still frames—single frames—placed on disc. There are at least four schools of thought on what is the fewest number of still frames that can be laid down safely. Some people insist that at least two frames should be put on the pre-master to assure that two correctly dominant fields can be found by the disc player. Similarly, others argue that a three-frame "landing pad" will assure that one frame (consisting of two interlaced fields) will be correct in field dominance when read out by the player. A still more conservative strategy holds that the computer in a Level 3 system may be otherwise occupied before or just after a frame search; therefore, the safest course of action would be to place four frames on the disc for each still that is needed.

I believe that one frame is enough. But that frame must be in the right place. And that one frame absolutely must be generated with correct field dominance. Having worked on nearly 40 discs, more than half of them computer-driven, I have never experienced a problem because there was just a single still frame in use. And when you are creating a consumer product or a point-of-information display, it may simply make the program unworkable to stack multiple stills on the disc when one will do. Bob Stein, of the Voyager Company, a deep thinker on the subject of videodisc, compares the medium to a book wherein still frames are analogous to pages. The philosophical structure of the disc does indeed parallel the dynamic of a book, but imagine thumbing through a book with three or four copies of each page. Annoying, yes? The same is true of disc.

No rule is invariant, to be sure. The peculiarities of your application or delivery system may warrant multiple copies of each still frame. But, in general, a single frame will do. And don't believe the myth that single frames are unreliable or unfindable. Every frame on a videodisc has an address and every frame is findable. If you have done your homework on field dominance and have been careful in post-production, you will have nothing to fear in your final product. And, your users may thank you for your economy of gesture.

Placing Single Frames

The problem of placing single frames on disc has to do with the videotape medium used to create the pre-master. Most editing systems, and most 1-inch videotape machines, have big problems in playing back a single frame and in certifying its precise frame location. Using a CMX-340X edit controller, the way to verify that the frame location is correct has often been to perform a preview of the next one-frame edit. To make the one frame visible, the edit operator inserts a test signal of black. You then watch a blur of still frames scream by (at 30 frames a second), followed by a burst of black which represents the desired location for the next still frame. This arcane process has sometimes been the

"The Metaphor
that Comes to
Mind is that One
Simply Nails the
Required
Number of Still
Frames into
Place and is on
One's Way. This
is Not True."

only way I could check the proposed location of a one-frame input. It can confer a sense of confidence, with practice—it is remarkable what we see at this blurring speed, but this procedure is never guaranteed to make your day.

Once the still frames are placed in their correct frame locations (hopefully) and your 1-inch pre-master is completed, it is often not possible to simply play back your tape to check the stills' locations. As you begin to jog forward on the 1-inch, carefully counting two phase bars or bands of static-like information in each frame (these represent the two fields making up the frame), at some point, your tape machine will start putting out inaccurate SMPTE frame location data. The tape physically shifts against the heads, and this microscopic shift is enough to throw off the accuracy of the time code. If your carefully crafted stills seem to have moved, just ask the edit operator to make a visibly time-coded dub of the pre-master, and then check the frame locations on the dub, offline. (For more information about post-production processes for IVD see Chapter 8.)

Once again, the terrible invisibility of videodisc prior to disc replication comes home to roost. It is truly maddening to have to use bursts of black to gather clues as to whether or not your stills are in the right place. And, facing the limits of $100,000 tape machines in playback does little to inspire confidence in the neophyte disc producer. But be brave, all of us live within the same limits. When we record directly on disc in a few years, we won't have to go through these gyrations. No doubt these trials only make us better people.

We have only spoken here of placing the still frame on tape. Let us consider preparations of the still frame material itself. In some cases, you will be inserting slides. These may have materials that have critical color information. In this case, color correction may be in order and each slide will have to be read through a film chain, then color corrected at the master switcher and only then laid down on tape in its proper location. This can be an exceedingly time consuming process, and a costly one. The National Gallery of Art disc required this sort of treatment. In fact, it required many of the slides to be reshot, since the colors had faded.

TEXT FRAMES

Most often, the still frames that one uses for training or point-of-purchase applications are text frames. These can be created on art cards and then shot by camera or created by an electronic character generator. Art cards are a perilous course of action since several layers of text resolution can be lost in the process. If the camera is in any way imperfect, blurring of letters can result, and the final message may be unreadable. It may appear more costly at the start, but in most cases it is more efficient to use a video character generator to convey a text message.

You cannot put as many words on a video text screen as you may like. You certainly cannot transfer a full 8-1/2 inch x 11 inch text page to a single screen. Practically speaking, you are usually limited to a maximum of about 40 words per text screen. The screen is 40 characters wide. The maximum number of lines of text is about eight, with six being a safer number.

A vitally important rule to remember is that character generator operators do make mistakes. You must be meticulous in checking every word of text. Few things are more embarrassing than wild misspellings on an expensive videodisc. Here, again, the tools for proofing the work are manual, and in your hands.

Every text frame must be formatted. That is, you (or the character generator operator) must choose the font for the lettering, the size and colors of the letters, the arrangement of the print, the color(s) of the background and the presence or absence of underlines and other markers. In truth, the limits of innovation in this connection are quite real. Red is a poor choice for background color, since it is usually accompanied by video "noise"; blue backgrounds can aid in imaging text; and the best colors for inside lettering are white and yellow. Watch out that your yellow doesn't tend too far to the green. Good colors for backgrounds also include pale magenta, light (lime) green, cyan, and warm but unsaturated orange. Browns and dark greens usually look horrible.

Obviously, there is no upper limit to text size. Be reasonable. As for lower limits: it is usually difficult to image print from a page (with an art stand camera) and then key or matte it to enhance its legibility—the edges subtly blur. The results of transferring print from photocopied sheets or book pages is often disappointing. Character generation is the best and safest course for dealing with text. The lower limit on letter size is a subjective matter. The disc will work for you since it will conserve your master tape's resolution—but you must keep in mind the kind and size of video display in the final user environment when you make the ultimate determination on how small you dare to go. In a pinch, I have squeezed as many as 16 lines of text onto a single screen—but I don't recommend it. Go for eight, and don't allow your client to browbeat you into an untenable position. The best way to avoid impossible assignments is to make certain that production or post-production people are involved in text frame design from its inception. Otherwise, well-meaning but uninformed instructional designers or computer folk may paint your still frame producer into a corner, having arrived at rigid (but unreasonable) requirements for numbers of words per frame.

PHOTOGRAPHIC STILL FRAMES

For archival, medical and training applications, you may be required to place on disc one or more sets of still frames based on photographic materials. The photos may be supplied as prints or slides. Slides are easier to deal with. Anytime you must use an art stand to image materials, the process of alignment and focusing will slow down the recording process. Slide transfer to tape through a telecine projector or to a filmstrip is mechanically simpler and therefore faster.

The process begins with the organization and coding of input materials. The slides should be cleaned and checked, examined in particular for physical blemishes or color inaccuracies. This is tedious work and it is wise to hire people who are both compulsive and interested in the content.

If the disc is intended to operate as a Level 1 product, be certain that the still images are transferred in the order in which you propose to organize your index, or the order in which you wish the user to encounter them. If the disc is to be part of a Level 3 system, or part of a visual database, your options are more numerous. You may wish to order the materials by format—starting with 35mm slides, then 4-inch x 5-inch transparencies, and so on.

The images must be coded and then listed, with an accompanying description of the content of each. When they are known, frame numbers can be added to this list. This information belongs in the production book.

After cleaning, check the images to make certain that they correspond to the order of the list. Then, check them again.

Methods of Transfer

The methods for transfer, on the way to the pre-master, are numerous. You can elect to transfer the stills first to film and later to tape. In this case, select 16mm or 35mm for the interim recording. Or, if tape is the target medium, you can use either a direct connection of camera to tape or the dedicated device known as a telecine projector.

Furthermore, you may decide to store the stills in an electronic still store (ESS), which can hold up to 2000 still frames, rather than on videotape. Digital video recorders, which can record and manipulate single video frames (Abekas A-64, for example) are available in many studios. The advantage of using a "slide-file" type of digital still store is that you gain some flexibility to reorder your still files prior to final down-loading to the pre-master videotape. There are cost, quality and flexibility benefits to each of these strategies.

Probably the fastest method—though not necessarily the least expensive—is to use a film-chain (telecine projector) feeding its output to an electronic still store. Laying down single frames on tape still entails an edit operation for each still stored; you will be limited by the pre-roll time on the videotape recorder in the speed of this operation. With an ESS, the limiting speed is how fast your operator can push the button.

In practice, people sometimes use a combination of telecine and art stand for inputting stills. In such instances, you simply record on a sub-master videotape, say, 30 frames (one second) of each still, inspect it later, and then, after you have checked the image, dump it on your pre-master as one or more frames in precisely chosen locations.

Automated transfer of slides to tape can cost as little as one dollar per image (or slightly less). Color-corrected slides cropped or framed through a master switcher can cost 10 times that amount. You pretty much get what you pay for.

After the images have been transferred, check the pre-master tape or a check disc on a frame by frame basis against your master list. When you are certain it is perfect, you can proceed to press the disc.

STILL FRAME AUDIO

To complement the vast visual information storage capacity of the videodisc, explorers in the field have sought a way to extend audio playing time beyond the 60 minutes of audio (on two 30-minute tracks) on each side of a disc—two hours on the most recent machines (with two-additional digital audio tracks). To achieve sound over still pictures, audiotape was tried but had obvious disadvantages in speed, durability and synchronicity. Compact disc has appeal, but it does introduce the expense of additional disc mastering to the project.

In 1983, Sony introduced a system that stored digitally compressed audio in the active video lines on disc. As the disc played, the compressed audio sequences were read into a RAM buffer within a still-frame audio decoder box. The box subsequently acted as a compressed-audio expander as it played back the stored audio in analog real-time. In 1984, EECO, Inc. introduced an analog system that offered a 300 to 1 audio compression ratio. Since then, Pioneer Video introduced its still with sound/data (SWSD) system and LaserData introduced its TRIO systems. These systems range in audio storage capacity from 1.2 seconds per video frame to 10 seconds of audio per frame—depending on the hardware selected and audio quality desired. The selectable audio quality grades were: (1) telephone quality, (2) AM radio quality, and (3) superior-to-AM-radio quality.

A disadvantage of these technologies was that the screen blanked (or blinked) while the audio information was loaded into the decoder. First, the compressed audio was accessed, the screen went black, the audio information was read into the decoder's RAM in video real time, video information was accessed and displayed, and the audio playback began. Blanking time varied with the system used. In the Pioneer system, only a portion of the video frame was lost, or masked—the top and bottom ninth. This area itself could be hidden by clever graphic overlay design. With the other systems, the technique was most effective in slide-show-type applications, in which the blanking did not call attention to itself.

Encoding for sound-over-still has been performed at various sites specific to each system—Tokorozawa, Japan, for Pioneer; Atsugi, Japan, and Sony Technology Center for Sony; Compact, Vidtronics and Windsor Total Video for EECO.

All of this is basically history since none of these approaches ever became widely accepted as a product. All were expensive—multi-thousand dollar solutions. And all were too complicated. Sony has basically bowed out, and no longer actively markets a product. EECO was sold, and goes on. But, for mortal producers, sound-over-still on a single videodisc remains too expensive to be practical.

If you are interested in pursuing sound-over-still, consider using a videodisc player with a frame buffer. Load in a still, then seek out the associated audio on either the two analog or two digital sound tracks. Such a feat is possible on a Pioneer 8000 (and may be do-able under bar-code control). Alternatively, use CD-based audio, accessed by the same computer that's running your Level 3 videodisc system. This may be the simplest path to quick, responsive and seamless sound-over-still. If you are interested in exploring this technology, get an updated list of manufacturers and contact them directly when you are ready to move. This is a technology in rapid motion, and prices, performance and product availability will change.

One dictum will always remain *au courrant* in sound-over-still work: it requires a perfect passion for detail. If you think field dominance is barely visible, check out digitized sound squeezed into your video. You have to review each recording immediately after it has been laid down, verify every location and keep perfect paperwork before, during and after the online sessions. But if your passion is information, analog-to-digital conversion, and the compression it makes possible, may be irresistible.

CONCLUSION

An interactive video producer faces endless details, raying alternative production strategies and layers of personnel. But, as an IVD producer, you must look beyond the "dits" to the experience all of this will create. Never forget the product. Don't lose sight of the user, who will, in truth, complete your creative effort. All of this is for naught if your final product plays best as a linear (and non-interactive) presentation.

8 Post-Production on the Pre-Master

The editing process for product intended for release on videodisc is similar to other video editing processes. It begins with an offline that yields an edit decision list (EDL) and concludes in an online edit suite. There are, however, a few critical differences that must not be ignored:

- For videodisc, always use non-drop-frame SMPTE time code when recording in the field and in the edit suite.
- The pre-master videotape has to be precisely formatted according to the guidelines of your intended mastering facility (Pioneer, 3-M, Sony, Laser Video, etc.). Check your intended mastering plant's printed guidelines and follow them to the letter.
- The overall length of the product cannot exceed the total amount of real estate on the disc: 30 minutes (CAV) or 60 minutes (CLV) on laser disc, 60 on VHD.
- Avoid placing footage or stills requiring high resolution within the first 1000 frames of a CAV disc. This particular area of real estate has lower resolution.

Remember that IVD is not a linear medium. Sections that need to be accessed most frequently should be near the middle of the disc. Frequently accessed stills (such as indexes, menus, etc.) should be clustered in this favorable position. Motion sequences (such as remedial segments) that may be accessed frequently should also be appropriately positioned on the videodisc.

The following is a list of the basic technical requirements for successful disc post-production:

- Pre-master videotapes should conform to broadcast quality NTSC specifications.
- The sync standard that provides complete information regarding match-frame edits and timing requirements is called EIA RS-170A (horizontal, vertical and

blanking sync pulse locations and durations).This standard must be carefully observed.

- The color burst reference signal must be present even on black and white programs.
- Black levels and contrast ranges should be uniform throughout materials intended for each disc side.
- Peak chrominance and luminance levels must not exceed 100 IRE units for highly saturated red or magenta fields; blues, greens, yellows and cyans must not exceed 105 IRE.
- Audio tracks should have a signal-to-noise ratio of at least 55 db, less than 1% total harmonic distortion, and no hiss or buzz. Dolby noise reduction may be used, but be certain you process the sound consistently. You cannot mix Dolby A and Dolby B on your source audio. (Most disc mastering houses require Dolby A.)
- Observe SMPTE safe picture and title areas so that captions and graphics will be visible on all televisions.
- Avoid excessive contrast and chroma levels, especially on text frames. (Disc mastering firms recommend that chroma levels not exceed burst.)
- In most cases, the head of the pre-master tape will require two minutes of color bars and audio reference tone (twice the normal amount), then one minute of black; the tail of the pre-master has a minute of black.

IN THE OFFLINE SUITE

The offline edit of a disc proceeds very much like any linear video edit—except that both editor and producer must remain cognizant of overall program length, relative positioning of segments, and conservation of real estate for still frame insertion. In the offline stage, the client is signing off on the content and sequence of information. The offline is several generations down from the master, has visible time code, degraded audio and no special effects, but it contains the exact content (minus still frames) of the final product.

During the edit of a disc I was producing in Japan for JVC, my Japanese colleagues felt that a "rough cut" should include still frames. This no doubt reflects their dedicated professionalism and creative zeal, as well as the strength of the yen. However, it is virtually impossible to produce a set of still frames and then insert them, quite precisely, into their intended location on a rough cut, since most offline editing systems cannot perform single frame edits. Therefore, I do not believe that it is necessary or prudent to execute all the still frames in the offline edit phase. Rather, it makes sense to mock up all the still and text frames, presenting each on paper or art cards, and to execute a handful of test frames on videotape to check legibility, background colors and overall formatting.

The offline edit is dedicated to working out the arrangement of materials on the disc, organizing the flow of linear segments, making certain that the relative positioning of materials makes sense—and, most importantly, to giving the client a chance to see and approve something before you go and burn the big bucks of an online edit.

THE ONLINE EDIT

The full complement of stills and text frames are added in the online. They are actually produced, in an electronic sense, in the online suite. Still images may have to be transferred from slides to an electronic still store, and then downloaded to your pre-master tape. Text frames are often created by a character generator or art stand camera and then massaged by the switcher (which can produce borders and background colors).

When you go to the online, this is what you should bring:

- The field master videotapes;
- The edit decision list (see Figure 8.1);
- The offline edit tape with visible time code (in case you forgot to copy something in making the EDL, it is wise to have the source in hand);
- All artwork, slides and other inputs for titles, stills and text frames;
- Complete documentation, including scripts, script notes, still frame design specifications, and text frame scripts.

At $250 and more (sometimes much more) an hour, you will want to have everything with you that can speed up the process.

The facility providing the equipment usually provides an editor as well. This person knows the peculiarities of the house equipment and is generally pretty fast at keying in strings of numbers representing the in and out points of each edit. All that the disc production team can supply at this stage is a post-production supervisor (usually the same person as the offline editor), who sits near the facility's editor and reads out the numbers for each operation. The post-production supervisor also checks the quality of each edit, both technically and aesthetically.

Customarily, the facility editor is quite expert in performing all sorts of tricky edits and will offer creative suggestions, when appropriate, in order to improve the quality of the product. These professionals know how to stay out of your way when you need to move quickly, but they also have a great deal of experience that can subtly, but importantly, contribute to your product's quality. Most editors listen a good deal more than they speak, and know when to keep their great idea to themselves. If you ask for their input, it will be offered. If you're departing from industry standards, or making a glaring error, the editor will speak up. Otherwise, the project's post-production supervisor "runs the edit," calling each shot, setting the pace, and generally acting as the creative mind in the process.

Other project personnel can be present at the online edit. But all input to the process should be routed through the post-production supervisor. It is usually inefficient, and sometimes simply disruptive, to call out your good idea from the back of the suite. Walk

Figure 8.1: Edit Decision List

Page ___:___

Edit Decision List

Title _____ Project No._____

Producer/Director _____ Asst. Director _____

Editing Technician _____

Editing Schedule _____
                     (date)

1 Event No.	2 Reel No.	3 Edit A/V	4 Special Effects	5 In Time Hrs. Min. Sec. Frm.	6 Out Time Hrs. Min. Sec. Frm.	7 Notes

up to your person, whisper in his or her ear, and they will relay your words to the editor. This is quite analogous to protocol on the set. There, all input is funneled through the director. Here, the post-production supervisor is the director of the edit. Deal with him or her accordingly. An online suite is no place for clients and no place for a creative jam session. That will slow down the work.

Timing

An edit day can run 10, 12, or more hours. Anticipate that there will be overtime charges. People should pace themselves. Tempers can fray in this pressure cooker atmosphere. Here, as on the set, egomaniacs are the kiss of death.

Each edit process has its own pace. Assembling a linear motion segment can take very little time. But building a complex digital effect for an opening can absorb an hour or more. Text frames and stills range in their time "cost"—allow yourself six to ten minutes per frame, after text has been entered into the character generator or the still image has been recorded on tape. Frame-accurate, field-specific edits should be checked assiduously as you go. Single-frame insert edits intended to correct errors are a tricky bit of business; it's better to get it right and then proceed. If you are entering the text at the edit session, don't plan to get more than four to six frames recorded in an hour. If everything has been carefully planned and preprogrammed, you may get 20 recorded in an hour.

Audio takes time. Just inserting music at the opening and close of each linear segment and sprinkling a few effects in appropriate spots can consume hours of editing time. Budget accordingly.

I suggest pre-building complex audio sequences offline and then inserting whole sequences in the online session. For example, a complex mix of sync sound, voiceover and sound effects can be arranged on the two audio channels of the rough cut and later simply mixed in the online. Rather than having to find each of the separate audio inputs, they will all be on one tape, in one place, readily accessible and mixable in the final session. To prepare such an input tape may not even require a mixer, and only minimum attention to sound levels. The key is to gather your materials in the most efficient way and then let the editor massage and position them properly on the pre-master.

CUE INSERTION

In some cases it may be necessary to insert videodisc address codes at certain disc locations prior to mastering. These cues are electrical impulses that tell disc mastering equipment where to lay down address codes on specific lines of the vertical interval between frames on the disc. These address codes will be read by the player. They may supply frame numbers, or they may tell locations of picture or chapter stops.

In the mastering process, a frame address code is entered on the master disc beginning with the first dominant field, and then on every other field that passes. Every frame on the master disc has a 40-bit frame number code on line 10 and line 273 in the vertical interval. This 40-bit code is read by Pioneer players of the 7820 series, and thereafter, and tells such machines which frame they are reading. A 24-bit code, entered on lines 17 and 18 or 280 and 281 of the vertical interval of the master disc, tell Sony, Magnavox, Sylvania, Pioneer and DVA 8210 players which frame they are reading.

Disc mastering equipment can be switched to read either cue information or composite video information, from which it can derive frame addresses. The mode used must be followed consistently throughout the mastering process.

Mastering is accomplished in real time. It cannot be interrupted until the full disc side is completed and all information has been read off the pre-master tape. Afterthoughts at the mastering stage are just that.

MASTERING AUDIO TRACKS

Pioneer Video Processing Services accepts separate audio tracks on 1/2-inch four-track analog audiotape or Sony PCM 1610 digital audio on 3/4-inch U-Matic cassettes. Formatting of a 1/2-inch four-track analog tape, recording at 15 ips, is recommended in this fashion by track number:

Track 1: stereo left channel (or mono) audio.
Track 2: stereo right channel (or mono) audio.
Track 3: blank (or 59.94-Hz resolver tone—if mastering facility is to conform audio to video).
Track 4: continuous SMPTE time code with offset specified from 1-inch C; program start should be at 1:00:00; beginning of tones should be at 0:57:30. (Non-drop-frame time code is required for Pioneer pressing at this time.)

It is required that time code extend one minute in front of and one minute after all recorded video or audio on the tape, including tones. The noise reduction system generally preferred is Dolby A.

1/2-Inch Masters

For 1/2-inch four-track audio masters the following tone sequence is necessary at the head of each reel for proper level setup, equalization, and head alignment. Have your audio person record tones with Dolby off.

For 15 ips:
0:57:00	1 kHz at 1 dB VU
0:57:30	10 kHz at 0 dB VU
0:58:00	100 Hz at 0 dB VU

0:58:30	Dolby operating level
0:59:00	One minute of silence
1:00:00	Beginning of active program
x:xx:xx	One minute of silence following active program

For 7-1/2 ips:

0:57:00	1 kHz at 0 dB VU
0:57:30	10 kHz at 0 dB VU
0:58:00	100 Hz at 0 dB VU
0:58:30	Dolby operating level
0:59:00	One minute of silence
1:00:00	Beginning of active program
x:xx:xx	One minute of silence following active program

Pioneer's pressing establishment allows pink noise or other tones, and they prefer that audiotapes be wound onto the take-up reel with tails out. Reels that are encoded with Dolby A should be so marked.

1-Inch and 2-Inch Masters

For audio recorded on 2-inch helical, quad, or 1-inch Type C NTSC videotape:

Video:	Up to 60 minutes of active program (including logos, bumpers, FBI copyright warning) per CLV side. Up to 30 minutes of active program per CAV side.
Audio 1:	Stereo left channel (or mono) audio. Both stereo audio tracks must be in phase throughout.
Audio 2:	Stereo right channel (or mono) audio.
Audio 3:	Continuous SMPTE time code. Non-drop-frame time code is required. At the present time, Pioneer will record continuous time code (for a price) before producing edit masters for disc production. Tones and bars should begin at 0:57:30. First frame of program video at 1:00:00:01.
Control Track:	No discontinuities.

You can't rely on your disc mastering facility to do anything with your audio other than laying tracks down and moving them in the most rudimentary way.

ADDING FILM AND SLIDE MATERIALS

Current recommendations for film materials include at least 12 feet of passive leader, followed by SMPTE Universal Leader with frame 27 punched for sound sync and frame 171 synchronized with audio beep. (This synchronizing information on frames 27 and 171 is required only if separate magnetic audiotape is provided.) Designate your film frame 219 as videodisc frame 0 and film frame 220 as videodisc frame 1.

Either negative or positive films are acceptable. It will help if there are as few splices as possible, and at least 12 feet of tail attached.

For 16mm film, emulsion should be on the same side throughout. This side should also be identified (emulsion wound in or out).

Film shot at 30 fps (frames per second) eliminates some complications related to 3-2 pulldown in film-to-video transfers. Since each film frame is transferred to two video fields in a consistent one-to-two pattern, picture cues are not strictly needed on the tape for full-frame identification. But, unless you control field dominance, you may have a problem with "flicker."

You can avoid flicker in tape segments originally shot in film at 30 fps if you answer these questions early on in post:

1. On which field (field one or field two) will the tape be edited?
2. On which video field (field one or field two) will the flying spot scanner begin a new film frame? (This establishes the field dominance of the telecine.)

A computer-controlled pin-registered film gate on the telecine (such as the Steadifilm gate accessory to the Rank Cintel Mark III) makes it easier to achieve smooth, steady transfers, and allows possibilities of single-frame video recording when certain master tape recorders are used in conjunction with the telecine.

Transferring slides to master tape can be done during the online edit, or added to the video transfer through a hookup with a digital still store. Similarly, single frames of text stored in a character generator may be added during the transfer.

The most important thing to consider is field dominance; if field dominance changes during part of the program, you may have to add in picture cues throughout the entire tape to tell the disc-mastering equipment where the changes occur.

With regard to sound recorded on film materials, either optical or magnetic audio can be used. On 16mm audio the outer track is preferred, but the inner track is okay. A 0-VU reference tone should be added, ideally in the film leader. If what you have is a separate mag sound track, either full coat or striped will do. Include 12 feet of passive leader. Pioneer recommends that a hole be punched in the separate mag to correspond with SMPTE Universal Leader frame 27 and that an audio beep be added to correspond with frame 171 on the SMPTE Universal Leader.

Again, add the 0 VU reference tone on the leader, and add at least 12 feet of tail.

PROBLEMS IN POST

Inconsistent Field Dominance

One game disc I worked on involved about 100 10-second linear sequences that were to be run by a computer board installed in an arcade game console. Problems arose when the videodisc came back from the pressing plant and was sent off to the computer programmer. When he tried to use the first and last frames of the motion video as freeze points—the interval in which the user was asked to drop another quarter—the programmer discovered that about 50 percent of the time, the image flickered hideously between two very different scenes. It turned out that the editor had set the videotape recorder on which our product had been pre-mastered on "auto," and the machine randomly selected field 1 or field 2 for the start and stop point of each edit segment depending upon its brightness and color values. The result, for us, was terror. Would we have to re-edit the tape and re-master the disc?

As it turned out, since we weren't using still frames, and we only needed relatively stable start and stop images, the programmer was able to find acceptable in and out frames. He keyed these start and stop points into the controlling chips' memories, and we were saved. Unfortunately for the user, the screen would occasionally have to go blank as the system searched for the next acceptable start point. The only alternative would have been repressing—and an extra cost of $5,000 (between editing and pressing).

Audio Headaches

In the category of audio headaches, I had an occasion on which an engineer in the field had mixed a Dolby sound recording with a non-Dolby sound recording. This produced some extremely interesting effects in the online edit suite, as the decoder desperately tried to make sense of this melange of audio processing techniques. The result was some rather odd final audio—rescued with extravagant equalizing efforts, but still oddly hollow and weird—and a fair amount (about three hours) of expensive lost time.

MISSING SEGMENTS

One of the lovelier scenes of the first MysteryDisc never made it to disc. In fact, it never made it to the edit suite. Apparently, this lovely tidbit, shot in the palatial gardens of the Vanderbilt Museum on Long Island, NY, somehow got erased during the original recording. Imagine our surprise and embarrassment on finding that it didn't exist.

The answer, it turned out, was clever editing. We discovered that a scene we had already laid down would never be seen by people viewing the particular segment that was missing. That is, if a user chose the storyline pathway that led through the missing segment, they would not get to see the already edited sequence. So, we simply recycled the wide shot, laid down another (very slightly modified) bit of voiceover narration by one of the characters, and collected four awards for our efforts. Whew! Moral: don't give up, and stay calm. Remember, producing is constant problem-solving.

Digital Eccentricities

D-2 is the hot component digital way to record your pre-master. It'll go a zillion generations and never show noise. Unfortunately, this digital advance in quality has a few attendant dysfunctional attributes. On D-2, a search backwards or forwards usually results in a blank screen. Anyone used to videotape editing is accustomed to orientating him- or herself to the images that blur by at faster than normal speed. To verify that you made the right move or just to check your progress, seeing the pre-master as it jogs is a valuable fact, lost when you work in D-2. Also, in reverse search mode, D-2 plays back a hideous audio tone; no information, just a blast of noise. Again, experienced producers and post-production supervisors, weaned on analog video, will find none of the usual orientation cues here. To verify that a track is down, or to find one's place, it has always been handy to hear the peculiar Bulgarian that is analog reverse audio. Well, it's gone on digital videotape recording. And that is not an advance, for the producer.

PRODUCING FOR VHD

JVC's VHD (very high density) disc system uses a contact reader. VHD first was offered to buyers in the United States in 1980, and was quickly withdrawn. Videotape simply swamped it, and JVC had a major stake in tape technology. But times have changed, and VHD in combination with the MSX II computer offers a fairly potent training system for less than half the cost of its closest competitor. That, JVC has been willing to bet, is going to interest somebody.

Currently, laser is beating VHD in popular acceptance in Japan, by more than two to one (and accelerating). A few years ago, VHD had the lead. The future is of course not certain, but VHD will probably be around for a few years and may enjoy some acceptance in the industrial training market. For this reason, a few words on its peculiarities.

The primary advantage of VHD is a significant cost advantage over laser in replication. Functionally, laser disc and VHD are quite similar. VHD provides for normal linear playback, still playback, variable motion in forward and reverse, stereo audio and discrete channel playback, and random accessibility to specific information on the disc. The most important differences between VHD and laser videodisc is that VHD revolves at 900 rpm (laser revolves at 1800 rpm), so VHD displays two frames per revolution, compared to laser's one. There are several implications of this difference.

A single rotation of a VHD disc yields what is called a "page," consisting of two frames. Slow motion is achieved by alternating play and stop (i.e., freeze) modes at the end of each revolution, while the disc continues to rotate at a constant 900 rpm. Quick motion playback is achieved by jumping and kicking the sensor at each revolution.

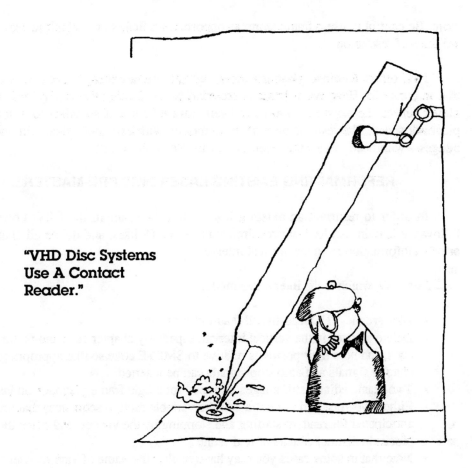

"VHD Disc Systems
Use A Contact
Reader."

Up to 99 chapters can be encoded on each side of a VHD disc. Chapters can be used to segment a linear program or to break up lessons in an interactive training program. Chapter numbers are encoded onto the disc during pre-mastering. An auto stop code freezes the motion when it is encountered. As many of these as are desired can be encoded on the disc.

In a stunning breakthrough, VHD offers worldwide standards compatibility. Any VHD disc, recorded in any of the three basic television standards, can be played by any VHD player, which senses the differences and modifies its playback rotation speed.

As with laser disc production, non-drop-frame SMPTE time code is required, continuous from the start of video. Luminence levels must be kept below 100 IRE unites, chroma at about 75%. But here's a key difference: the edit point in pre-mastering must be at the first field of the next even numbered frame. Moreover, no chapter can be less than 10 seconds in length.

The two-frame constraint of VHD requires that still pictures and materials designed to be played back as slow motion be recorded twice. This can be accomplished using normal linear editing techniques, reiterated, or through application of an electronic still store. Be careful to use a frame store, as opposed to a field store, which retains only half the lines of resolution.

JVC offers a unique playback mode termed "extra editing" (because it takes just that to create it). Here every frame is recorded twice. During linear playback, the VHD player senses the segments that have been extra edited and switches to twice normal playback speed. The result is normal speed motion with extremely clear still and motion images—with nearly twice the resolution available on laser disc.

REFORMATTING EXISTING LASER DISC PRE-MASTERS

In order to reformat an existing laser disc pre-master in the CLV mode, simply follow the lead-in and lead-out requirements for VHD discs, and delete all chapter cues or other information from the vertical interval.

For CAV, which is the interactive mode:

- Observe the VHD lead-in and lead-out protocols.
- Delete all cues in the vertical interval, especially chapter cues and picture stop cues, and note their positions relative to SMPTE code so that appropriate VHD chapter signals and auto stop signals can be inserted.
- Two-frame edit all still images designed for single frame playback on laser disc.
- Either two-frame edit or extra edit the variable motion sequences that are anticipated for random starting and stopping by the viewer, and place these blocks in distinct chunks of real estate.
- Note that in some cases you may have to alter the name of various user functions such as "step" and "still" presented in overlaid text frames, in order to be consistent with VHD remote control keypad functions.

One last special feature offered by VHD is the so-called "multi-lane system." Here, a digital field memory device is used to lay down successive fields (half frames) of up to four distinct programs. During playback, any one of the four fields can be selected by kick control of the sensor, under external computer control. The sensor tracks the program path, and upon command directs the sensor to a different part of the same track. The effect is of an instantaneous jump into a different video sequence. Whether for role-play or for simulated road racing, this offers the ultimate in instant responsiveness to user performance.

CONCLUSION

Now you are prepared to edit what you've produced for IVD. By following the guidelines above and by adhering to the printed pre-master requirements of your disc presser (3M, Pioneer, Sony, et al), you should have few problems. Adherence to pre-mastering standards, attention to field dominance, care about real estate allocation and proper documentation will insure your survival and prosperity in this critical phase.

Pioneer's Disc Pressing Facility In Carson, CA, Utilizes Elaborate Automated Disc Replication Technology In the Manufacturing of Laser Discs.

Photo courtesy of Pioneer LaserDisc Corp.

Inside a Clean Room, a Contamination-free Environment at Pioneer's Disc Pressing Facility in Carson, CA, a Matrix Operator Prepares the Glass to be Used in the Mastering Process of Laser Discs.

Photo courtesy of Pioneer LaserDisc Corp.

A Robot Arm Simultaneously Loads and Unloads Laser Discs in the Metallizing Process, Where the Reflective Surface of the Disc Is Created.

Photo courtesy of PioneerLaserdisc Corp.

Part 3
The Future of
Interactive Video

9 The Continuing Development of Disc

After a decade of struggle, the interactive videodisc has arrived. The virginity of American business in connection with IVD is a memory. People know a bit about this technology—so it is no longer an impossibly futuristic sell. Indeed, growing familiarity with the plethora of interactive technologies is giving businesses pause before they commit resources to any one.

Visiting the LaserActive Conference in Boston a couple of years back (one of a growing number of shows that deal with IVD), I was struck by two things: the size of the turnout and the fact that nobody there was overweight. This led me to conclude that, after years of doldrums, there was real hope for laser videodisc as an industry; but, as before, only the lean would survive.

These days, you can choose from a bouquet of varied interactive video conferences. The largest is SALT, the Society for Applied Learning Technology, held in the summer. Growing in prominence are the Multimedia Expos, presented twice a year by American Expositions. There is the heavy-hitting CD-ROM Conference in the spring, sponsored by Microsoft, which features all the new developments on the frontiers of visual computing. The Intertainment Conference in the fall gathers the interactive entertainment industry, a growing coterie. There is considerable interactive hoopla at MacWorld. The CES and COMDEX shows feature all consumer and industrial disc hardware. In the 1990s, it should be relatively easy to find a conference featuring interactive video—delivered in one form or another—just about every month. Consult *The Videodisc Monitor* for their detailed calendar of events.

ALTERNATIVES TO VIDEODISC

The biggest news of this hour is the immanence of both compact disc-interactive

111

(CD-I) and digital video interactive (DVI). Both are about to be introduced as real products. Both promise full-motion video under real-time interactive control, delivered on a CD-ROM (digital) disc. One, DVI, looks like the ultimate in training and decision support/presentation tools. The other, CD-I, bids to be the biggest splash in consumer electronics in our time.

COMPACT DISC-INTERACTIVE

CD-I is a close relative of compact disc (CD) audio. A little background on compact disc may be useful to producers who hope to create interactive multimedia products for the consumer. This is the realm that beckons those seeking the future—computer visuals, retrieved from small discs and massaged by ever more capable chips. For producers who wish to avoid obsolescence, there is no safety in the analog video domain.

The CD offers a choice of formats for data of different kinds: CD-digital audio (CD-DA) for high-quality audio playback, CD-video (CD-V) for brief video clips, CD-read only memory (CD-ROM) for large quantities of digital data, and CD-interactive (CD-I), described by its developers (Philips) as "the first fully interactive combination of sound and pictures, computer text and graphics on one system." The following is a brief typology of the varieties of CD:

- CD-DA audio discs measure 12 cm across, and carry up to 72 minutes of top-quality digital audio per side.
- CD-V discs range from 12 cm "singles" with six minutes of video and 20 minutes of audio to 20 cm and 30 cm (8-inch and 12-inch) discs offering 20 minutes and one hour of video per side, respectively.
- CD-ROM discs are essentially computer storage media, holding up to 600 megabytes of data on a 12 cm disc—the equivalent of 150,000 pages of text.
- CD-I will hold up to 650 megabytes of data on a 12 cm disc, and can handle a variety of media, including video still frames (up to 7800), audio (between two hours of high-quality sound and 17 hours of narration), text and graphics (150,000 pages' worth) or, more typically, a combination of these under control of a computer program, also stored on the disc.

It is important to note that CD-I is not a form of video, but a digital, computer-based technology. Sound and images are aspects of a databank. The key to understanding CD-I, and to producing for it, is to think of it as a computer process, not as a television or textual process. In this, as in other CD-based technologies, the key constraints have to do with the rate at which data can be transferred from the disc, the amounts of data required for different elements of a CD-I program, and the memory space available on the disc and in the controlling system.

Key questions in this medium are:

- How much space is required to store each type of data?

- How fast and to what locations can data be moved before decoding and playback?
- How much of the available memory and data channels are occupied by each effect?
- What happens on the screen while data is being sought elsewhere on the disc?
- How many tasks can be handled by the main processor?

Basic Operating Principles

Data on a CD-I disc is stored on spiral tracks of sequentially recorded "sectors." Each sector contains just over two kilobytes of data. The CD-I player reads data at a constant rate of 75 sectors per second, or approximately 170 kilobytes per second.

The information capacity of the disc is consumed according to the level of audio quality, image resolution and coding techniques used. Images range in their requirements from a full 170 kilobytes for a single RGB still to 85 kilobytes for so-called DYUV images in normal NTSC resolution, to as little as 42 kilobytes for CLUT graphics. Blocky animation, produced by so-called run-length coding, can yield an entire screen of images with only 8 kilobytes of data. The requirements of each coding technique are quite detailed (and can be read in the *CD-I: Designer's Overview*, a publication of Philips International). The key point, from the producer's perspective, is that information bandwidth is a resource akin to the disposition of real estate on an analog videodisc. The producer must make the rate of use of information the very core of his or her design considerations. Audio quality, level of picture detail, rate of image update—these are decisive elements in the matrix of CD-I, and they are intimately interconnected.

A sequence of program sectors are grouped together to produce a program module. The designer breaks sequences into short modules that the user then couples together through choices made at the screen interface. The CD-I player reads one sector of the disc at a time and allocates it to a data transfer channel. There are 16 data channels available for audio, and 32 channels for other information. All of this information is interleaved, or sequentially intermingled, on the disc. A certain percentage of the data stream is occupied by audio, and the balance is dedicated to picture and control information. All pictures are transferred from disc to 512 kilobyte blocks of random access memory (RAM), before display on the screen. The time it takes to move an image from disc to RAM will be related to what other activities occupy portions of the data stream in that same interval. It takes about half a second to load a photo-quality image into RAM, if no audio is passing through the channel at the same time.

"Cognitive full motion" is where no subjective motion jitter or blur can be seen by 95% of the population. This requires a minimum of 10 frames per second. Only about 13% of the screen could be updated fast enough to achieve full-motion video using the chips available in 1989-90. The then-current trick to create motion on CD-I was to perform only partial screen updates or to use run-length coding (treating images in lines rather than pixels) to reduce the amount of information required to store an image. Software coding techniques have been developed that increase the amount of screen available for partial updates. Full-motion, full-screen capabilities have been demonstrated and are about to become CD-I state of the art. It is nonetheless useful to be mindful of the system's hardware constraints on data flow. Full-motion capabilities are supposed to be present in the "base-case" CD-I player, to be released in 1991. It could be years before this promise is kept.

Use of Picture Planes

There are two picture planes on the display screen, a foreground plane and a background plane. In addition, the cursor operates as a separate visual foreground element. This structure offers a rich variety of visual presentation schemes.

In order to minimize the amount of information updating required, a useful trick is to load a photo-realistic background image on one plane and then create a limited area of motion on the foreground plane. In the ABC Sports/Fathom Pictures golf title, for example, the fairway views are photographs from a Palm Springs golf course; the golf club wielding player, who is under user-control, is a foreground image taking up a fraction of the screen.

Various effects can be performed between picture planes, including dissolves and wipes. These effects are generated within the CD-I player and the images themselves remain untouched. Images can also be organized into montages and clustered on the screen.

A powerful use of the multiplane capability is to compose a graphic control panel, to be overlaid on the screen. The combination of images and text into a display panel can organize the user's control of a game or allow a learner flexible access to progressive segments of a training program. It can outline the locations that can be chosen in a mystery game or recap the clues that have already been uncovered in an interactive detective story. The control panel can be coded in a rapidly accessible form, and it can be summoned at random, throughout play.

In addition to the two pictorial levels, the designer can overlay text on the composite image, essentially as a third visual level. This adds enormous flexibility to visual design for entertainment or teaching applications.

The multilayered construction of pictures on CD-I challenges the interactive producer to design with, literally, new visual dimensions.

CD-I Product Design and Development

American Interactive Media (AIM), acting as the software development arm of Philips/PolyGram, requires prospective CD-I co-publishers to conform to a regimen of development and documentation. First, a program developer submits a brief description of his or her concept, with a verbal sketch of what the user's interactive experience might be. If this initial query elicits interest, the developer enters the formal program design and production process.

AIM actually pays for the development of proposals that attract its interest. In a 60 to 90 day period, a developer must come up with what AIM calls the Idea Map—which includes a detailed treatment of the product, preliminary flowcharts for the program, storyboards illustrating key interactions, possibly a *HyperCard* stack demonstrating the workings of the title, a list of design team members and a production budget and schedule. The storyboard documentation will be fully fleshed out as the project moves toward production. It may be desirable to mock up some motion sequences to illustrate the visual and auditory workings of the product as well, in the first phase.

Authoring and Production

The product then enters the authoring or production phase. There are several pathways being developed for creation of materials for CD-I. It would be wise to contact AIM directly to get an update on their recommended production methods for your intended product. At this writing, visual materials are being created on Amigas, Mac II's, Suns, and via component video for use on CD-I. Ultimately, all audio, textual, graphic and visual elements must be translated into a form that can run in the OS-9 operating environment of the CD-I coding equipment. The techniques for doing that were in flux for some time, although AIM asserts that working authoring platforms have been installed around the country to perform this final, crucial step in realization of product.

The production facility for executing CD-I multimedia product will consist of computer workstations, dedicated audio and video servers, a number of read/write discs, write once optical discs for archival storage and magnetic tape drives for output to the mastering plant.

The authoring system will have a role in each stage of production. In design and scripting, the designer must be sure that elements of the program will not exceed the limitations of the CD-I player. Disc bandwidth, memory limitations, seek times and other design constraints are analyzed to be certain that the project does not violate these boundaries.

After design parameters have been established, the authoring system guides the assembly process. This consists of:

- Data acquisition—the capturing, editing and encoding of audio and video.
- Presentation editing—linking audio and video with programming.
- Disc building—which combines control information with encoded audio, video, text and application programming.

All of these elements are structured hierarchically into records and files, and are processed into forms that are needed by the master tape generator.

In short, we interactive producers have come a long way from the analog translation of the 60-year-old television waveform into microscopic pits on an analog videodisc. Making CD-I discs leads the producer deep into the country of interactive digital multimedia—the new promised land of consumer electronics for entertainment and education.

Stretching Interactive Applications

CD-I will be sold with a number of user interfaces—ranging from joysticks to track-balls. It is a visual computer whose manufacturers don't want the consumer to know that they are buying a computer. So this machine will sneak into the consumer's home under the pretense that it is just a souped-up Nintendo that happens to play CD audio.

The real challenge in this new medium is, as always, to the designer and product creator. How frequent and how dramatic will be the opportunities for the user to control the action? What new kinds of experiences will be offered, using the mix of text, graphics, motion visuals and high-quality audio? How will information be linked inside products? Will we develop new control metaphors—beyond on-screen menus and Hypertext links? How flexible can we make the user's experience, and still capture a few couch potatoes?

The first products on CD-I will inevitably have the aura of repurposed artifacts of earlier media. But in time, producers will begin to create materials that make full use of the remarkable capabilities offered by multimedia computing. In some sense, prior to wide distribution of reliable authoring systems, the new forms are literally unthinkable. But with experience—with growth in both the creation and the use of these new products—a new vocabulary of multimedia experience will begin to develop. The designers of the 1990s will reach far beyond the experience IVD producers can today bring to bear on this new challenge. As before, the rule must be: Stretch. Have guts. Shoot for the stars. Our imaginations themselves will be expanded by the explorations we now begin.

DVI: DIGITAL VIDEO ON THE PC

In the professional arena, digital video interactive is making a hit. At the core of this technology lies an assessment of the perceptual characteristics of the video display and digital video compression.

DVI began its life about five years ago. Following its birth, it was very nearly overlooked in its home laboratory—the RCA David Sarnoff Research Center. It was not lost, luckily, and DVI made its first appearance at the 1987 CD-ROM Conference, stunning attendees with a feat that no one expected for another half a decade: full motion video running off a CD. The problem of compact disc had been summarized in the words: "It's like sipping the ocean through a straw." That is, the compact disc can store huge amounts of information. But the data transfer rate is pitifully slow—if you happen to want to see pictures. For audio, 170 kilobytes per second is fine, as we know. It is serviceable for still pictures as well. But for motion ... a big trick is needed.

Compression had been vital to the teleconferencing industry for years, and a good deal of science had gone into software compression algorithms before DVI was conceived. But still, to cram what broadcasters take five million cycles per second to deliver into a space 1/300th the size, data-wise, was no mean feat.

The heart of DVI technology is a pair of VLSI chips—a pixel processor and a display processor. These two chips contain more than a quarter million transistors apiece. They are programmable, and so can be given new instructions at every cycle of their operation. The result is more than facility at noting changes and handling pixel details; they can flexibly manipulate the very images they are compressing. This endows the user with an unprecedented power to shape the visual content of whatever appears on the screen.

Through the combination of high-speed processing chips, powerful compression algorithms and elegant microcode, the DVI hardware turns an IBM PC into a digital video image display system. There is a complex technology at work here. I commend the interested reader to Arch Luther's *Digital Video in the PC Environment* (Intertext Publications, McGraw-Hill, New York, 1989). From the perspective of the interactive producer, there are many key points to keep in mind, in order to optimize a DVI product.

Design

Commencing the design of a DVI production, thought must be given to the mix of media to be included in the final disc. What grade of audio, and how many minutes of it? How many still images, and what length of motion imagery is required? What sorts of transitions or effects are demanded? What range of capabilities will be conferred upon the user? These questions will be answered in the design document. And from the answers will come an assessment of the scope of the control code writing task and the range of production tasks facing the production team.

In the design phase, consider how the viewer is going to interact with the program. Note that some authoring systems do not allow you to interrupt a video segment. Check the proposed authoring system before production begins, studying its limitations and features. Then, design your script and your interactive experience so that the viewer does not get bored waiting to be asked to participate in the program.

Consider: How do you propose to fit video into the screen? DVI has the ability to display video in different areas as well as in different sized windows on the screen. Full screen, half or quarter screens are all possibilities that offer the designer creative opportunities. Remember that you can place your interactive buttons (i.e., controls or "hot" screen areas) almost anywhere you want, at any time.

Production

In production, original recordings should be made on the highest resolution format possible. DVI images get stepped on as they go through the digitizing process. Load the dice in your own favor, as much as possible.

Key Constraints in Shooting

Avoid fast pans or snap zooms. These are brutal challenges to the compression technology, and may result in boxy, aliased and highly unattractive images. Try to let the action flow inside a pretty steady frame. You can cut pretty freely—avoid high-speed montages. But normal, steady cuts, every five seconds or so, present no difficulty to the system. Opening and closing shots should be steady for a second or so, at the ends. This gives the chips time to get oriented before things begin changing.

Post-Production

Post-production of materials for DVI again offers deviations from the norm. Beware of fast dissolves. These are very tough for the DVI chips to make sense of. Take at least one second to execute an effect, to permit the chips to keep up with scene transformations. Pattern wipes hold up nicely. Keyed edges also translate very cleanly. The main concern you must have here is pacing. DVI gets by, but it cannot rock along at top speed. Give the system a chance and it can keep up with nearly every normal video trick. Plus, it offers your users a few of its own. Recall that users have a wide menu of digital video effects at their fingertips to manipulate DVI image files displayed on their screens.

Audio

Recorded voices and other sounds can be digitized in real time. This permits rapid progress in preparing audio files. Instant playback allows error detection and correction, at top speed. But here's a problematic feature of DVI in the audio domain. Let's say you want to present a series of stills with audio narration. DVI accesses information at different rates, depending on how information is stored and also depending on how large the files are. DVI stores information on a disc (an optical or magnetic disc) in clusters of file types—not by programming moments. This means that all the video clips are in one place and all the stills are in another. And all the data or text or audio information is in yet another place (these places being directories). So, when a bit of program requires a still image with text or audio overlaid, the system has to search all over the disc before it can display the requested information on the screen. This takes time, and you, as designer or producer, must plan for these timing realities.

Time Considerations

Speaking of time, it takes significantly longer to display a high resolution image off

CD-ROM than it does off a SCSI drive. Therefore, if you are trying to time an image to audio, you must provide for enough load-in time. Still images load at different rates, depending on the compression of the file. DVI compresses still images based upon textural information; so images of a complex and varied surface will compress less than an image of a simpler landscape.

A final item not to be overlooked: the so-called streamer tape that comes back from Intel's DVI-Princeton operation, after they have compressed your video (1-inch, Betacam or D2 are preferred formats), should be readable by the tape drive that you are using. Check on this. For DVI developers using Pro750 development systems, this is not a problem. For us mortals, who may join this bandwagon while it jostles a bit further down the road, take appropriate precautions.

Recall that you can do your own compression to perform in-house alpha tests on your intended product. For brave souls, Intel provides a convenient method of digitizing your videotape in your own shop. This is not the professional level video compression that DVI is justly renowned for, but it does offer a simple way to see what your program will look like with video clips as you begin to put it together. Playback will be at 11 to 15 frames per second (although audio will play back at normal speed). This so-called real-time video compression method has great value for checking out timing and sequence manipulation under interactive control. And it can save you valuable time while you await the return of your compressed video on streamer tape, for final product integration.

Authoring

All the disparate files we have considered—audio, video and instructions for interaction—are brought together on an authoring platform, in order to create the streaming digital tape mentioned above from which the CD-ROM will be mastered. This is the domain of authoring systems. CEIT Systems' *Authology Multimedia* is a leading software tool in this area, as is *MediaScript*, available from American Helix. These packages are constantly being improved and debugged. From my experience, I recommend that you take a capable C programmer along when you venture to this altitude. Solo climbers are not ready to tackle DVI's cloudy heights. Things are getting better; but it still isn't easy. Don't let anyone kid you that multimedia authoring is trivial. It is not. There are lots of simple, logical things that are hard to do. Things do not work exactly as you may think they will. And processes hang up. This business entails huge blocks of data, moving to the right place at the right time. Authoring tools are the container freight handling devices in this high-speed information storm. It will be a while before the tools are impeccable.

Still, we should be of good cheer. There are a few people around who have wandered in these precincts. We can always find one and employ him or her on a first DVI job. We're still in the Daniel Boone phase of digital multimedia. Live guides are worth their weight in gold.

Developing an Organizational Plan for DVI Production

In Arch Luther's book on DVI, he gives a summative example of a simple DVI production, by way of warning: he proposes an introduction to the technology, using video, followed by a graphics presentation about installation of DVI, a few demo programs, the complete text of his book and some hardware and software documentation of DVI, accessed via an index. Luther proceeds to show what a hairy task such an undertaking would be. Clearly, there are more steps to DVI production than in producing a Level 3 videodisc as suggested in the accompanying chart (see Figure 9.1), which is in truth too schematic. An awful lot of work goes into a DVI project. It seems that much more than careful planning and design, followed by meticulous production and editing, is necessary. Every step in DVI production entails a calculation of memory demands, data transfer rates and integration of codes and functionality. Neither attention to detail nor excellence in production can alone suffice to deliver an excellent product.

All relevant data must be collected, reviewed and sized in order to plan the application's contents. Implementation of the user interface must be specified at the outset. Then, each audio, video, textual and graphic production task requires a treatment of its own in order to plan the layout of contents, as well as the production enterprise. Scripting and production planning follow this detailed front-end analysis of the tasks ahead.

In Arch Luther's example, the data types that must be integrated in his proposed point-of-information application include:

- 1 megabyte of DVI software documentation
- half a megabyte of other text files
- 1 megabyte of C code examples
- 0.7 megabyte of book text
- 50 still photographs
- motion video and audio (demo)—50 megabytes (compressed)
- 7 more minutes of motion video—introduction and installation
- 1 megabyte of index files

Given this embarrassment of information riches, DVI production will no doubt foster the development of advanced production control and documentation tools. All of the various tasks must be hyper-linked. Each component's activities must be displayed and causally connected to all other sub-groups' production plans. In other words, the project manager needs to be put into a computer, so that everyone can proceed into this fertile new terrain in relative safety. Key steps are already being taken. *ScreenPlay* (from Vent) and *MediaMaker* (from MacroMind) will begin to empower producers for the work that is coming. Fear not. Study up. Do some reading. Go to a few trade shows. And we'll see you on the king's electronic highway—videoconferencing from desktop to desktop, using DVI, in 1993.*

---

*My gratitude to Philip Malkin, video and multimedia producer, for his valuable input on producing for DVI.

Figure 9.1: DVI Production Sequence

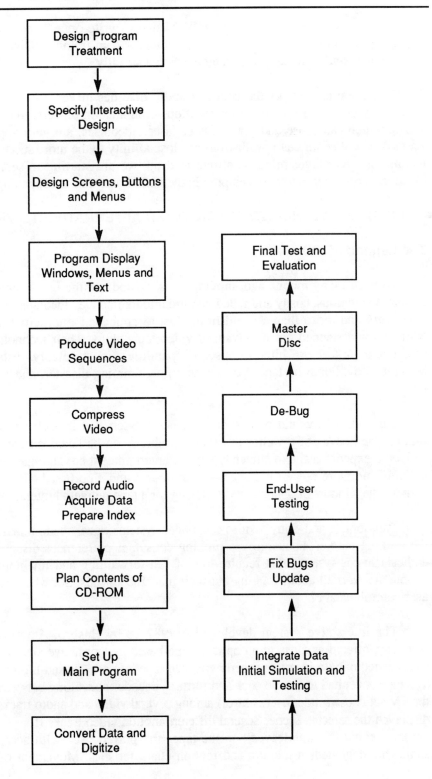

WORM DISCS AND COMBINATION PLAYERS

Elsewhere in the disc world, WORM discs (write once, read many times) are proliferating in document storage and retrieval, and in training. Disc applications in training and education are growing, modestly but steadily.

The great success of the compact audio disc has driven consciousness of laser technology deep into the public imagination. Combination players, like Pioneer's CLD series, which can sense and play CDs or laser videodiscs, are very appealing indeed. Laser disc will continue to profit from its close affinity to the more successful CD audio technology. As player prices continue to drop, we are starting to see, at last, a real videodisc consumer market developing in the U.S. Laser disc is happening.

INTERACTIVE VIDEOTAPE: ANOTHER ALTERNATIVE TO VIDEODISC

The Legend of ISIX

Not too many moons ago, moving vans cleared out the last remnants of ISIX— Hasbro's ambitious, costly and failed venture into interactive videotape technology. After two years and more than $20 million, Hasbro conceded interactive home video to Nintendo and stepped out of advanced video technology. Their technology worked; it was a *tour de force* of hitherto unseen capabilities (most notably, instant branching) brought to the home on tape. But it was a programming disaster, and thereby hangs a tale.

Right now, it would be perilous at best to assert that interactive videotape has a future. The hooves of a welter of optical products can be heard on the horizon. Chip-delivered experiences, like Nintendo's home entertainment box, are garnering billions of dollars. And more powerful devices—both optical and chip-based—are just around the corner. The window of opportunity is closing for a tape-delivered interactive medium.

Panasonic's and Sony's video responder systems never made much of a splash in the training market. Focus groups probing consumer preferences discovered a lack of enthusiasm for systems that require lots of fast-forwarding and rewinding of the tape. People are worried about wearing out their machine's heads. So, where is the future for user-controlled tape?

The interactive system developed in 1987-88 by Hasbro, America's largest toy company, offered as close to an optimal tape-based system as we are likely to see. The user needed only to insert a properly encoded tape into a VCR and hit the PLAY button. The tape never had to shuttle back and forth. A decoder box sitting between the VCR and the TV set chose (under user control) among several video and audio tracks and instantly displayed the selected scene. Several different motion video tracks (four turned out to be optimal) could be "interleaved" on the tape, and played back with acceptable, though somewhat degraded, resolution and motion characteristics. Movement of a joystick or pressure on a control button instantly signaled the decoder box of choices being made. The resulting motion video, taken from among the streaming parallel movies, would be selected and displayed in a disturbance-free fashion. A variety of graphics and text could be overlaid on the motion video, and multiple audio tracks (up to 16) could be instantly selected and presented as well.

This remarkable system offered most of the features of a Level 3 videodisc system at a bargain price: about $300 for the decoder box, $200 for a VCR and $300 for a color TV. Here was the long-sought after, but never seen, under-$1000 high-level interactive video system.

Elements of this technology will no doubt be seen again—in smart TVs, advanced interactive cable, and more capable VCRs.

Hasbro Backs Out

Alas, ISIX (as a computer named Hasbro's electronic subsidiary) bit the corporate dust in the fall of 1988. Inflated D-RAM costs pushed the price of the decoder box higher than Hasbro felt they could sell it (to $300 from $150)—far past the price level of a toy or an impulse buy. So Hasbro backed away from their $22 million investment in interactive entertainment.

In truth, failures in software production were at least as critical in killing the system as the momentarily inflated price of dynamic random access memory chips.

From the start of Hasbro's venture, massive errors in program production strategy, fiscal control and scheduling had undercut the enterprise. An $800,000 production was delivered for $2.2 million. The rights to *Police Academy* and *Star Trek* were secured for millions of dollars, and no usable product was fashioned from them. Schemes, never properly documented, were committed to production. To the end, there was never an agreed upon format for presenting designs, program flowcharts or scripts. A producer who executed one game's motion footage openly admitted that she had no idea how the game was played. This is not recommended procedure for interactive video production.

When the smoke cleared, about $10 million had been spent in creating barely three viable games. ISIX went down and another puff of smoke obscured the future of full-motion interactive video entertainment in the home.

A lot was forgiven along the way at ISIX because everything being tried was quite new. New, yes; unprecedented—no. The management of ISIX over-dramatized the boldness of their video software venture, and did not mobilize an abundance of prior art that might have informed their efforts.

The production budget, for example, has been with us for decades, as has the production schedule. And videodisc producers know a lot about documentation of complex, branching storylines—along with accompanying script formats, flowcharts and production designs.

ISIX acted as if it were pioneering interactive movies out of whole cloth, and they made, and re-made, every mistake there is to be made. The errors were compounded by a simultaneous commitment to not use union actors and to only use "name-brand" directors and known software titles.

Non-union actors do not belong to SAG or AFTRA for a reason—usually, lack of experience and work. And, Hollywood directors are perhaps the worst people to hire to create non-linear materials. They think linearly. They like to let the material "breathe," and they are wildly expensive. Hence, the spectacle of fabulously lit, expensively set material, that was amateurishly acted, and produced at staggering expense. In the end, one had to congratulate Hasbro corporate management on their ultimate good sense. But one might also rue the failure of the boldest venture yet, on American soil, into the realm of interactive movies. So ends the legend of ISIX.

While the ISIX project failed, the technology still remains, and there may be an afterlife for ISIX's "NEMO" hardware, so labelled because it was "Never Ever to be Mentioned Outside." A major Japanese chip maker, for example, could salvage the ASICs at the heart of the box and build an interleave-decoder capability into smart VCRs and TVs of the 1990s.

No matter who lifts this mantle, the trick remains to make intelligent and appropriate use of the multi-tracked interactive medium. What would YOU do if you could braid four stories together, and then pick'em apart, branching instantly among the strands?

Adventure tales come to mind, of course. So do a number of educational games, game shows, mystery stories, and user-controlled soaps. In truth, there is no end to the programming one can make with four tracks of motion picture, 16 audio channels, still frames and graphic overlays.

The one crucial producer constraint in this tape system, or any likely successor, is this: the medium runs forward in real time. All of the tricks of reaccessed video and audio we frequently mobilize on disc are of no use here. You can't go backward or forward to play a selected segment. You can grab and freeze a frame easily enough—and then await the arrival of the next still or linear segment you'd like to display. A judicious mix of sound and frozen image can do wonders to fool your audience into believing that they're seeing a motion scene. In one ISIX game, this trick was used frequently, and it works well. A freeze-frame of an empty room with crickets chirping outside reads as a real-time view of an empty room. But remember: all of your calculations about real estate and track switching must be conducted against a background of a continuously advancing real-time linear medium.

The best tool for laying out your design is a long wall and a stack of filing cards. Mark a minutes and seconds time scale along the top of the wall, and array the cards in four parallel horizontal lines, according to where each segment (or still) should fall. Allocate one file card to each motion segment or set of stills.

If audio is a vital dimension and it branches widely, add parallel file card tracks above or below the video channels that "house" the audio.

A Final Caveat

Computer game designers don't have all of the future in mind. Nor do linear video/film producers. The right approach to interactive program design and development lies somewhere between and beyond these two populaces' capabilities.

Real people will speak and move before cameras in recording the new interactive media. Control systems will receive coded instructions to play or freeze the picture. But those who look to just computer-video game creators or movie-makers for the entertainment and training software of the future may fail to deliver. These people speak different languages. They work in different ways—computer hackers do it alone, at night; movie-makers move and think in packs.

The future of multimedia programming will be shaped by a hybrid breed of producer and media manager. They should be:

- conversant with authoring packages and competent at line production;
- capable of flowcharting and sharply observant in the online suite;
- and undismayed by sound bites or megabytes.

Such a person would be useful in realizing the IVD production revolution. We shall have to nurture this new breed.

The Inner Workings of Interactive Tape

The ISIX product—regardless of its ultimate fate—promises an entirely new generation of television media products and services. These capabilities are bound to pop up in future generations of VCR and cable TV decoder equipment. The ISIX innovations may change the way home VCRs and even cable television are used.

The two basic elements of the system are:

1. A uniquely formatted TV signal that compresses multiple channels of video program material and digital control data into a single composite signal. This signal can be recorded on tape or distributed via cable TV.
2. A decoder box with one or more viewer control units. The decoder resolves the composite signal into distinct and separate channels. The control units enable the viewer to interact with the system in order to change "channels" (i.e., tiers) or to

input other responses or commands, depending upon hardware configuration and program instructions on the videotape. (See Figures 9.2, 9.3 and 9.4.)

In use with home VCRs, the concept of multiple "tier" or channel programmable action video creates radical new possibilities for live-action VCR video games, new forms of entertainment, training and education. Closed circuit TV can be made interactive, so that a viewer can enter responses in a training setting or suggest an outcome in a game show or entertainment piece. If a multi-tier signal is transmitted over cable, the viewer can select, at the control unit, which of the several tiers he or she wants displayed. For example, a viewer can select which of several camera views of a sporting event he or she wishes to see—and change this at will.

When a VCR reads a multi-tier encoded tape, the decoder receives the multi-channel signal, resolves it into the separate program tiers, processes the tiers according to instructions recorded on the tape or from the viewer's handset control unit, and outputs one final displayable signal to the TV set.

Fundamentally, the capability offered is much the same as switching channels on a four-channel TV set, except that in this interleaved system, the program tiers are related and coordinated within a single NTSC television signal.

One might, for example, create a video game in which a viewer is allowed to look at different aspects of the same event and then try to detect some pattern or detail within the event—as in solving a crime, or tracking down a lurking beast. Or, the multi-tier capability would allow a sports instructor to show the same technique or motion from several different angles. In education, the multiple tiers can be used to pose a question to the user or to prompt a multiple-choice answer and then switch to the tier that responds with analysis of the specific answer.

In any of these examples, it is possible to keep a record of tier selections made by the viewer at various points along the tape and to store this in memory. Then, the tape can be rewound and replayed, repeating the tier switches in the same sequence—on one's own machine or on a neighbor's. In this way, the viewer becomes a tape editor and assembles a unique program from the multiple tiers.

In addition to two audio channels per video tier, there are up to eight auxiliary audio tracks available for mixing in any way.

To author for this delivery system, a proprietary workstation is required. This consists of off-the-shelf VCR and editing equipment and a specially built encoder unit. The workstation is used to interleave the multiple programs that come out of post-production, along with auxiliary audio tracks and a digital data tier, all melded into a single NTSC signal.

The data tier contains title execution instructions for the home decoder box. These

Figure 9.2: Video Block Diagram

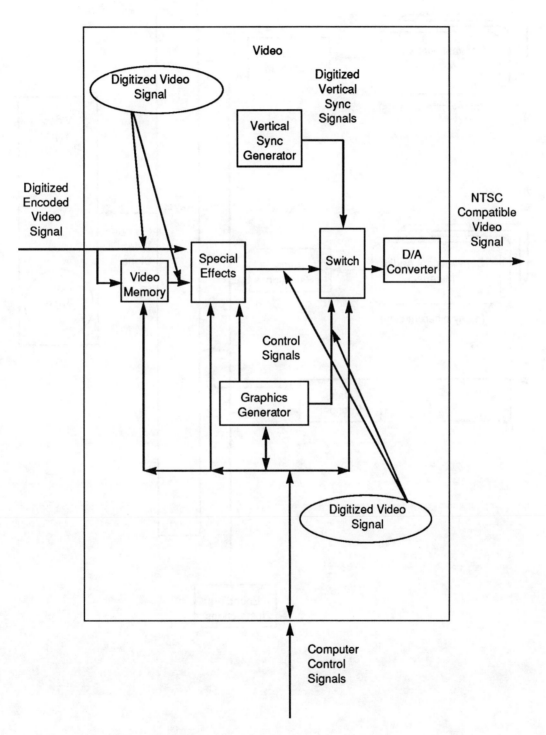

Figure 9.3: ISIX Encoding System

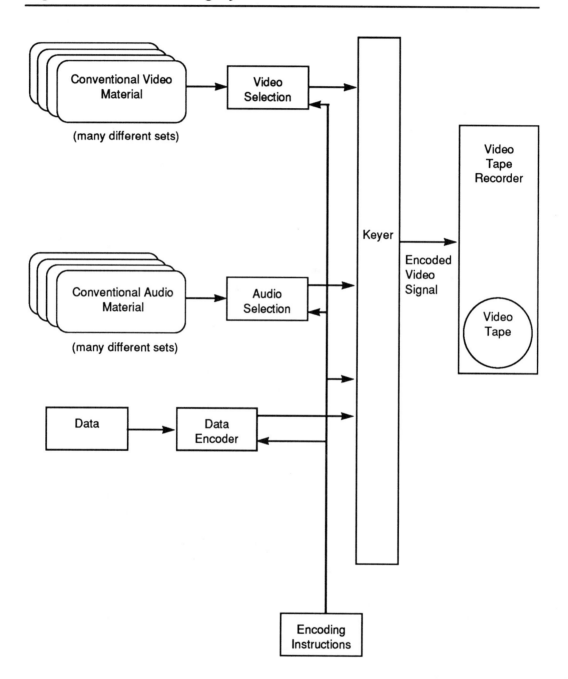

Figure 9.4: ISIX "Hologram" (Interactive Videotape) Processes

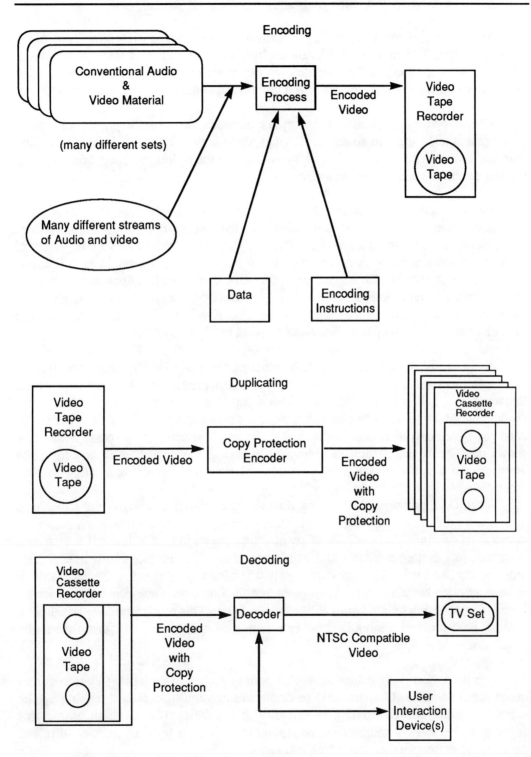

instructions indicate how the decoder must utilize the multiple video tiers and interact with the user's hand-control unit signals at each point along the program.

Production of multi-tiered programming is of course different from other teleproduction. The product is nonlinear. Instead of one story, there are several—mutually interdependent and interactive—and they are produced in parallel. All of the stories must be continuously synchronized in time and content.

There is an entire realm of interplay between tiers in storylines, character development and tempo. In all repeat viewings, the title must retain utility and value, and dramatic impact. The viewer is frequently involved in the action, dictating camera points of view and techniques of camera motion.

The interactive designer of the product has the same role here as in interactive videodisc production. He or she is the keeper of the concept of how the multiple story tiers interact to form the integral title. The designer brings the game's theory—genre, story, game mechanics (what control the viewer exercises over action on each tier, and game reaction to moves). During pre-production, the director and scriptwriter must work closely with the game designer so that before production, everybody understands all of the tiers and each of the scenes that must be shot, how the tiers relate, what the time tolerances are at decision points, and other relevant mechanics of the game.

As in videodisc, all scenes must be blocked more tightly with respect to timing, exits, entrances and camera movement, than in linear production. A time line for each tier must be established and integrated into a common master time line for the entire production. Multiple tiers often branch back to a single common point. It is here that the need for synchronization between tiers and the critical nature of the master time line shows up. The subject matter must develop in such a way that there is no conflict or unexpected discontinuity, should the viewer switch tiers at any given moment.

As in IVD, principal photography (or videography) has essentially the same steps as in linear production, but the process is more exacting. The master time line puts severe constraints on timing of individual scenes, and may require additional script-note personnel just to time entrances and exits from scenes and durations of action. Scenes must be executed precisely and improvisation is virtually eliminated. Because two or more tiers often branch to a common point and then diverge, there can be many match frames for the action to blend smoothly from tier to tier. This requires precise attention to camera position and angles, and a high degree of quality control throughout the production process.

If a production falls behind schedule, pulling scenes or collapsing them becomes much more complex. Changes tend to ripple through multiple tiers, breaking up the master time line and destroying the integrity of the entire title. Even an on-the-spot adjustment becomes a comprehensive replanning job by a team of people—director, game designer, scriptwriter and timing assistants.

The same constraints of multiple tiers and rigorous time lines carry through to post-production. The mechanics of videotape editing are the same as always, but they are more involved because multiple tiers must be cued to each other and synchronized according to the master time line. The editor must constantly be mindful of the other tiers because wherever the viewer switches from tier to tier, the action must be seamless. For action-oriented scenes, editing is the last opportunity to adjust the duration of the scene and fit it to the master time line.

In addition to the normal evaluation of story impact and continuity, the producer, director, writer and designer all become involved at various points throughout the editing process in the task of assessing game play.

Audio is more involved as well, since in addition to the two sound tracks for each tier there are eight auxiliary audio channels available for mixing into any of the tiers in any way.

In addition to video and audio post-production, there is the matter of integrating title execution instructions into the master tape. The title execution programmer has written instructions for the decoder box. These instructions take the form of a computer program which is encoded as running data along the tape, located in a multiple line header of each video field. This data is interpreted by the decoder to perform any tier switching required in the title, to interface with the viewer through the hand-control unit, and to keep game records.

The final recorded signal, delivered on tape, is a modified version of the standard NTSC broadcast television signal format. The video is encoded into the magnetic media of the videotape as a series of sequential diagonal tracks. In conventional tapes, each diagonal track represents one video field and successive tracks represent successive fields in the program. In this instance, successive tracks represent not successive fields of the same program but interleaved video fields from different program tiers. Digital data is integrated into the video field. The microprocessor inside the decoder box uses the digital control information to select a particular video program tier and associated audio and to mediate interaction with the viewer's hand-control unit.

The unique part of this video signal is the digital data header inserted in the first few lines of each field. During the invisible vertical retrace time, as each 262 1/2-line NTSC video field is painted on a TV screen, there is an available interval of ample length to insert several lines' worth of additional information—in this case, digital data. This invisible interval is termed the header. Here, one finds identification of the following field as a video or a data field and notation of which program tier this field is part of. Each field also gets a number, starting at the top of the tape, in a fashion reminiscent of videodisc.

This is how the ISIX interactive tape system worked. It is instructive because of its integration of video and digital control information, and because interleaving is a powerful tool for instant branching in training, simulation and entertainment. Whatever the fate of this particular system, we have only begun to see interleaved video with accompanying digital data.

STRANGE BEDFELLOWS

A very smart man, Steve Russell, inventor of the first video game and an important software engineer on the ISIX project, pointed out something that all of us who would meld computers and video need to keep in mind. There could not be a stranger or more incompatible combination of habits and orientations than one finds in computer games designers and in video producers. In games design, you hack and you change and you modify again until it feels right. There's just no way to know how it'll really come out until you get there. In video production, may the Great Spirit help you if you try that approach. Computer programmers who are hot simply don't like to do the same thing over and over. TV programmers like to do things that are in some ways familiar and predictable. They like to be pretty sure of the range of outcomes and of surprise. Not so computer hackers. Computer jockeys work alone, and they expect their products to be encountered by people who are highly distracted. Distraction (and multi-tasking) is, in truth, the name of the game. Video folk, in a nod at their filmic ancestry, expect people to sit still and pay attention when their wares are shown. Well, games are simply not experienced that way.

So, the result of these divergent world-views is much angst when the two teams must collaborate. These people simply cannot see the world through their opposite number's eyes. Often the result (at best) is a very high degree of anxiety throughout the production process, and a lot of discomfort on the set. Extraneous anxieties in production settings are the kiss of death. And yet, here they are. And how can we hope to avoid them, given the utterly different creative styles and Weltanschaungs of these two mutually dependent groups?

Increasingly, in interactive teleproduction, we find ourselves acculturating computer and production folk to the other's concerns. Computer people are rarely movie-creators. And video program creators are even more rarely truly adept at computer manipulation. The broad gulf that separates these populations is the domain of interactive media that many of us hope to inhabit, someday. If we do not bridge these differences of habit and thought, what will become of the future—which lies between computers and television? Probably the answer is to simply accrue experience. We do not need theories of how we can meld our world-views as much as we need the experience that will weave together our cognitive styles and capabilities. Still, the next five or ten years will be bruising ones for programmers in both media, as video and computing come closer and closer to one another.

INTERACTIVE VIDEODISC MARKETS

Despite the many alternatives to IVD that have been developed, the videodisc market in both the United States and Japan is growing.

The U.S. Consumer Market

The time for a major marketing move in promotion of consumer videodisc in the U.S. is here. The Japanese have been developing catalogs of interactive products that could justify adding another device to the pile of electronics in numerous American living rooms today. Movies, available for rental on videocassette at $1.99 per title, will never sell disc machines in America. There has to be something unavailable on any other medium that is sold (or rented) on disc. The answer is, as it has always been: interactive, user-controlled programming. Unless and until there is an extensive menu of software available on disc—musical, games, adventure, current affairs—there will be no sudden upsurge in popular acceptance of videodisc.

At one time, the theory was that IVD's time would come when every house had a computer (a goal that is stalled if not receding, certainly not coming closer). Then, as a massive visual database peripheral, the disc would come into its own. To be sure, IVD has lots to offer as a computer peripheral. But this confluence is simply not happening today. To integrate a computer with a laser disc player can be a troublesome affair, requiring hardware from multiple vendors. It can be done, and is being done in lots of places—but it costs some money, sometimes takes debugging, and is a few years from being a consumer product. (In truth, all it requires is an RS-232 port on the disc player, a cable and the right software drivers in the PC.)

The primary candidates for driving disc technology into the marketplace have so far failed to deliver—the arcades, the museums and the schools. Educational uses of disc are only now beginning to take off. There are some brave aspirations—Texas hopes to create a landmark science education curriculum on disc, to go into space-age science classrooms they plan to build. The Texas authorities have judged videodiscs to be textbook materials and so purchasable with those funds. Optical Data Corp. is suddenly a very rich videodisc school curriculum provider. California has some money for educational technology innovation. By and large, for most of the past decade, the IVD ball has been in industry's court. With the advent of video computing platforms, industry's eyes have shifted to PC-delivered multimedia. But, the biggest current player in the videodisc scene is the military.

Military Use of IVD

In 1987-88, the Pentagon began to buy more than 30,000 computer-driven videodisc systems. This so-called "EIDS procurement" was, up to that moment, the biggest thing ever to hit videodisc. (EIDS stands for "Electronic Instructional Delivery System.") Literally billions of dollars are being spent to feed the maw of this monster network. Everything from tank maintenance to language instruction to soft-skills management training is slowly but surely being translated from paper to disc. By the late 1990s, the U.S. military hopes to have a lot of spare file cabinets to sell. The ambitions

**Strange Bedfellows:
Computer
Programmers
and Video Producers,
Co-Creaters of
Multimedia Titles.**

are great, but actual spending has at various moments been modest. But, stay tuned. The *Business Commerce Daily* and *The Videodisc Monitor* carry EIDS procurement notices. EIDS is big, and it is going to contribute to the evolution of interactive video—simply through the mass of experimentation it is spawning.

Japanese Software Innovation

Whatever happens with EIDS, the videodisc marches on. Pioneer LaserDisc Corp. is making money. Yamaha and Sony are doing well on discs in Japan, and even Matsushita, corporate parent of VHD, is now in the laser disc column. In America, the growth of the laser disc consumer market has brought large laser disc distributors such as Image Entertainment to the point where they are willing to fund software development—proof of the existence of a healthy market. There are lots of reasons for optimism, if one takes a long-range view.

An interesting paradox concerning disc's future entails an important role reversal for the Japanese. The Japanese have not had a reputation for software innovation. But they are behind pioneering initiatives in program development and production for the disc. Apparently, the Japanese have figured out what RCA never did: you gotta have eggs before you can sell chickens; you must have software to entice people to your hardware. So, Laser Disc Corporation is making some of the moves the disc field has not seen since the disappearance of Optical Programming Associates—they are funding highly innovative programs for the consumer market. A variety of New Age music has been set to beautiful nature imagery to create music-image pieces. The Pioneer Artists jazz catalog is extensive. The great cities of the world are being packaged for disc. The best in music clips, graphic commercials and special effects artistry are catalogued on disc. Soon there will be periodicals customized for presentation on IVD.

Pushed forward by the Japanese, who are taking the first steps in this hitherto unfamiliar role of software developers, we are seeing some significant growth in the U.S. consumer market for disc players and disc software. Good news is visible: more and more video stores stock discs for sale. Discs happen to be perfect for rental, since they don't wear as do tapes. Videodisc player owners buy, on average, 30 times as many discs per machine per year than do their tape counterparts.

All that's missing is the software. Why are you just sitting there?

EPILOGUE TO THE FUTURE

There are important changes afoot. Apple appears to be scaling back its many bold ventures in optical media. Intel controls DVI, and is working with IBM on that critical piece of video computing's future. CD-I is going to come to market in 1991 with a couple of dozen titles and a player, for under $1,000, intending to reshape the consumer's relationship with computers and visuals. Commodore is releasing its "Baby," called CDTV, for under $1000, with a lot of repurposed Amiga software and a few new titles. And videodiscs are selling, well, into schools and homes.

From a producer's perspective, things look promising, but are not yet perfect. The games market is still dormant, both in the home and the arcade. LaserDisc Corporation of America's consolidated operations on the West Coast have yet to bear much software fruit. At the turn of the decade, it looked as if the Japanese had given up hope on a U.S. consumer market for interactive videodisc. But now sales of players are cooking. And both LDC and Image Entertainment are considering funding IVD titles for a renaissant home market. DVI is still a couple of years away, and offers scant opportunity for software producers. CD-I is approaching launch, has had a bumpy software ride, but claims to have working authoring systems and full-motion capabilities, as well as some promising titles in development. If AIM could get moving, CD-I might even flourish as an interactive publishing channel.

As to the rest of the interactive industry, the point-of-purchase system area is creeping along, EIDS grows slowly and steadily, interactive tape is dead, teleconferencing is flourishing and Big Blue (IBM) is starting to make big movies in multimedia.

A veritable alphabet soup of technological possibilities sometimes obscures the horizon: CD-I, DVI, CD-ROM, CDV, LV-ROM, CDTV—the list of would-be optical disc delivery systems seems endless. And all of these analog, digital or hybrid delivery modes have their proponents—powerful, rich proponents. Whither, then, interactive video? In the 1990s, which will be the winner? And to the producer, does it matter?

From a producer's perspective, it makes no difference how the images are delivered. The fundamental design and creative control tasks of the interactive video producer remain similar. But the opportunity to practice one's art, and the magnitude of one's income, will reflect the rightness of reckoning the winner in the great delivery system sweepstakes in the sky.

Affordable Erasable Discs

Probably the most important development in optical disc will be the advent of recordability—affordable erasable discs. The price of OMDR disc machines has declined by about half in five years. The $20,000-plus Panasonic disc recorder of 1985 is available for $12,500 today (the TQ-2026). Teac's OMDR player can be bought for well under $4,000. This trend will continue, and new technologies for erasable discs are proliferating. When disc recorders cost $1,500 and blank reusable discs cost $20, we will be in a new ball game. Stay tuned, 1993 could be a really big year for IVD.

Meanwhile, CD-I is well funded, but as yet has a limited supply of righteous software. Some smart people still won't bet on this horse. CD-ROM looks like it may harbor good game and learning system opportunities. But the key to CD-ROM is hot compression schemes. For now, the prize may go to DVI. Then again, Intel is no powerhouse of marketing. Neither is Commodore. Whatever they do, when IBM lays the digital video interactive chip set into the millionth PS/2 PC, in 1992, we will be in new terrain. When people become used to seeing motion video coming off their computer screens, we will truly be in a brave new world of training, gaming and learning.

10 Melding Computers With Video at the Apple Multimedia Lab

CASUAL MULTIMEDIA

At the Apple Multimedia Lab, the technologies of the computer and video have worked together synergetically for a few years. The Lab is now part of Apple R&D, operating as a unit of the Advanced Technology Group. Product development per se ceased in 1990. A small staff has been left to do theoretical as well as product development support work, but the lessons of the Multimedia Lab live on.*

Apple Computer was built upon a foundation of educational applications. But only since 1985, with the Macintosh computer and desktop publishing, did Apple seek to seriously penetrate the business market. With the development of optical media—CD-ROM alongside laser videodisc—Apple foresaw the development of business and educational applications of computing that involved text, graphics, sound and video. This mix of media offered radical new opportunities for the presentation of ideas—in the

---

*The Apple Multimedia Lab continues to exist, but it no longer has a charter to evangelize multimedia or to bring new technologies to the schools. The Lab was once part of Apple Computer Marketing. Today, operating under the Advanced Technology Group, it is part of the product development process, and therefore its work is more confidential in nature. Its overall mission is to assist Apple in bringing multimedia products to market, according to its business manager. It is no longer a high-profile promoter of the virtues of the new media. But still, the Lab remains a design-centered enterprise and its personnel—a smaller staff—continue to do evaluation and field study. On occasion, this brings them back to the schools. But this Apple Multimedia Lab is a fundamentally different creature than the one founded in 1987.

"The Mix of Media at the Multimedia Lab Offered Radical New Opportunities for the Presentation of Ideas."

classroom and the boardroom.With its friendly user interface, Apple felt well positioned to exploit the promise of multimedia computing.

In 1987, to probe this mix of media, and their interface with computers, Apple established an off-site laboratory, staffed by video producers and educational technologists. Their mission was to probe and push back the limits of multiple-media applications, driven by computers, for use primarily in the educational world. This was the Apple Multimedia Lab. (See Appendix to Chapter 10 on page 143 for an in-depth interview at the Apple Multimedia Lab.)

By early 1989, the Lab could report some significant findings. They had defined a mission that would make their work accessible to many who would follow. The Multimedia Lab specifically tried to avoid super-high-ended applications feats that would daunt any who tried to apply the lessons. Instead, they spoke of "design examples"—interactive multimedia prototypes that addressed specific communications challenges—as their goal. The objective was, throughout, to be accessible in tools used and in interactive product created.

The Multimedia Lab's staff was convinced that a lot more discs had to be fashioned before educators and office workers could know how best to create and use their own. The industry suffered from a shortage of viable design examples. In fact, people had trouble talking about their concepts for interactive projects, because, particularly in the domain of multimedia production, the product is just too complicated to be described. You have to have seen one.

As the Apple folk surveyed this somewhat tortured landscape, they became convinced that optical media were, in their term, "design-driven technologies." That is, it was in the design dimension that the most crucial contribution was made. If a series of compelling examples of multimedia productions could be created, this would offer more than features in the landscape; it could provide a trail for others to follow. If a series of affordable tools for shaping these multimedia productions could be specified, this would open the road to a flood of followers.

"A major assumption of the Multimedia Lab is that the interactive medium should be available for fluid manipulation and multimedia composition by everyone," affirmed a 1989 year-end report. The goal was to include low-end users such as teachers, students and business professionals in the process of multimedia production—as well as high-end users such as software developers and publishers. "At a multimedia workstation, people can construct original interactive presentations by cutting and pasting their own images and sounds with those already provided," the report went on to explain.

Here was the concept of hypertext writ large, in the realm of all media: to unite things, in the fashion of a creative mind; to wind together strands of knowledge, and present many sources at once upon a single screen.

With what equipment did one fashion "casual multimedia" productions as Apple set out to do? Apple's Lab made extensive use of a write-once videodisc recorder. This videodisc was controlled by a Macintosh computer running *HyperCard*. Any text, graphics and control structure that was needed was incorporated into *HyperCard* stacks. Additional images and audio could be digitized into the computer, and later called up by *HyperCard* as required.

The OMDR disc player (and blank discs) were still a little pricey. So the Apple folk often got a quick-press LaserVision videodisc, playable on any laser disc player. The quick disc and its accompanying control software could represent a completed "design example." If more than five copies were required for distribution, a normal replicate disc was ordered.

To input still images, the Lab used an inexpensive film-chain—a camera, prism and projector arrangement that permitted video recording of slides. The Sony ProMavica offered an alternative means of capturing video stills in a photographic setting. (We'll surely hear more of that technology in the 1990s.) To record original motion footage for a design, the Multimedia Lab used an 8mm Sony camcorder. The lab utilized 1/2-inch

VHS and 3/4-inch tape as well, but 8mm had virtues of extreme light weight, extended light sensitivity and high audio quality. Many of the Lab's projects were based upon existing film footage.

An Apple Scanner with *HyperScan* software allowed images to be scanned directly into *HyperCard*, with control over size, orientation and gray scale. The scanned images could be modified using the painting tools provided with *HyperCard* or other software. Video frames were digitized using Pixelogic's *ProViz* Video Scanner. Koala's *MacVision* also allowed images from a video camera to be digitized. Sounds were digitized using Farallon Computing's MacRecorder digitizer and *SoundEdit* software. All of these elements were then accessed by *HyperCard*. *HyperCard* was used as both the control and display core of Apple's multimedia systems. Because of its flexibility, *HyperCard* supports many kinds of educational projects. The *HyperCard* screens ('cards') can be used to present text, drawings, animations, images and sounds. Links between text, images and sounds can be created easily. The cards can control the videodisc segments and provide access to individual images and sounds on the disc. Because of these and other virtues, *HyperCard* significantly speeded up the design process and indeed opened the design and production activity to people not skilled in computer programming.

"The Potential Results of a Casual Liason with Multimedia Are Not Necessarily the Most Favorable."

Videodisc player drivers could be added to the *HyperCard* software packaged with the Mac by Apple. These drivers were placed in the *HyperCard* home card to allow the user to create buttons and other icons that accessed videodisc frames and other player operations from *HyperCard*. The *HyperCard Videodisc Toolkit*, available from the Apple Programmers and Developers Association, provides videodisc player drivers for many standard videodisc players.

The Apple Multimedia Lab used Crawford Communications in Atlanta, GA, to press their single-copy LaserVision discs. These cost about $300 apiece, and are made on plastic (as opposed to glass). Twenty-four hour turn-around is available. Two sides can be pressed for about twice the cost.

Table 10.1: Equipment at the Apple Multimedia Lab

Equipment	Approximate Cost
35mm slide-to-video transfer system	
Tamron Fotovix	$3,000
Sony ProMavica System Camera—MVCA7AF	$4,000
Edit/recorder/player	
MVR-5500	$3,800
MVR-AA770	$3,000
Sony 8mm Handycam CCD-V9	$1,600
VHS 1/2-inch camcorder	$1,500
Sony ProWalkman audio recorder	$ 370
Sony U-Matic (3/4-inch) edit system	
VO-5800 player	$5,500
O-5850 recorder	$8,500
RM-440 edit controller	$2,100
8mm Sony EVO-720 edit system	$4,900
VHS edit system	$2,500
For-A time base corrector	$4,000
Apple Scanner	$1,800
Pixelogic ProViz Video Scanner (color)	$1,700
Koala-MacVision	$ 350
Farallon Audio Digitizer	$ 200
MacroMind Software	
VideoWorks II	$ 300
VideoWorks HyperCard Driver	$ 100
Authorware—Course of Action	$ 695
Macintosh Plus	$1,800
Mac SE (with 20 megabyte hard disk)	$3,900
Mac II (with video card)	$6,800
color monitor	$1,000

Crawford works from a non-drop-frame time-coded 1-inch master videotape. The company can work from any videotape source, but if you send in 8mm, or VHS, or an uncoded 3/4-inch tape, there is an additional $200 charge for producing the 1-inch pre-master they require to press the disc. Non-drop frame time code must be continuous from the start of video through the program's end.

Table 10.1 lists some of the equipment used at the Apple Multimedia Lab. Apple Multimedia's dream was that this not-too-terrifying list of gear could find its way into many high schools and colleges, and that slowly but surely, many people would get their hands on interactive computer-driven videodisc. These explorers would be inspired, in part, by the small but growing number of "design examples" in the field. And a critical mass of understanding, dispersed experimentation and evolving applications would help lead us all to the promised land of interactive multimedia. Say, Hallelujah.

But the entire notion of "casual multimedia," to this bruised survivor at least, seems an oxymoron. Laser videodisc production is hard enough. Those who know say that CD-I projects make videodiscs feel like child's play. Apple's Lab may have trained horsemen to oppose tanks. There may be no casual survivors in the complex multimedia environment of the 1990s; at least, that's what I fear.

There are many interesting and important things that one can attempt on disc without a heavy heart or a full wallet. But it may be dangerously misleading to lure the uninitiated to this costly, invisible and increasingly complex medium under the pretext that it is fun, and that product can ever be delivered easily, unguardedly, casually. Videodisc production is more the province of Sun Tzu (author of *The Art of War*) than it is the domain of Lao Tzu (author of *The Way of Life*).

APPENDIX TO CHAPTER 10: INTERVIEW AT THE APPLE MULTIMEDIA LAB

The following is the transcript of an interview conducted by Marty Perlmutter (MP) with Kristina Hooper (KH), Director of Apple's Multimedia Lab. The topic of the interview is hypermedia and multimedia.

Kristina Hooper remains the director of the Apple Multimedia Lab, though the Lab operates differently today than it did at the time of the interview. Ms. Hooper's design philosophy and the conceptual roots of the Lab remain unchanged, so we present this interview for its insights into Apple Computer's deeper designs and multimedia strategies.

MP: I guess the first step is to get your definition—a real definition, not a theory definition—of hypermedia and/or multimedia.

KH: There are a lot of words being used to describe the same thing. Some of those words include hypermedia, hyper TV (a new one), multimedia and interactive multimedia. We have two kinds of multimedia—multiple media, such as books and records, which includes a little bit of everything; and multiple media such as sights and sounds. And we have new media which includes hypertext. We have hyper words, media words, we have all the words. (Laughter)

A few years ago, it was right to call this department the Multimedia Lab because it distinguished us within a computer company for doing things that were sound and image intensive. Now we are involved in developing a new computer medium. It's not, in my opinion, a videodisc medium, but a way in which computing grows. The new object for us now is the videodisc. We add live action footage and moving images to our repertoire. Whether a video signal or some other kind of signal is used, doesn't matter. For now, videodiscs are terrific sources of moving images and still images, so we use them a lot. It's a new way to bring information into a computing environment.

Given we're in a computing environment, we build imaging tools that let people make new things. If we were a movie-making company, which we are not, we would talk about taking movie images and adding interactivity to them. So people in the movie business will talk about multimedia computing or video computing. That's coming in from their direction. It's like taking movie-making and adding a little bit of computers. From the moviemaker's point of view, it's a matter of taking the primary piece, which is the visual element, and giving a little more control to it. It depends on what view of the world you hold.

Ultimately, in my opinion, there will be this "thing" called the information age, in which you won't talk about the technology. For example, you don't talk about computers, you don't talk about movie projectors, you don't even talk about the movies. It's information, and it appears in many forms.

The new opportunities appearing now are lots of sights and sounds. And these optical technologies are great for delivering them.

MP: So the mission of the Lab is to bridge existing technologies, and to use these as imagining tools?

KH: That's right. And using these imagining tools, we make things we call design examples. A design example is something that shows you what an actual event would be like. If we focus on the human experience, for example, what's it going to be like? The design example shows you what learning DNA's code would be like. Or it shows you what learning about the Constitution of the United States would be like. Or it shows you what a sales brochure or a business presentation would be like. And our hope is, as we show these possibilities, that we can drive the technologies. (That's what is meant by "design-driven technologies.") Then we get the size of memory, or the graphic overlay or the sound compression we need. That's the way we're affecting the technology to accomplish what's in our imagination.

MP: So you're an applications lab. Your reports go back to Apple Headquarters, and say, "We need 600 more megabytes of memory in order to reach this goal," or, "What you're not noticing in the sound area is this, which we learned in one of our rough simulations."

KH: That's right.

MP: One of your designers constantly says, "I don't care about the hardware!" And he means it, because it would, in a way, corrupt your mission.

KH: Yes.

MP: And yet, in the end, you do care passionately about the hardware, don't you?

KH: Yes. Yes.

MP: The whole point, the reason money gets poured into this place, and information gets poured back, is to inform the hardware effort, correct?

KH: That's right. What we don't do is maximize for a videodisc presentation. You see, if you're a videodisc producer, you will maximize to make sure you get your quick jump or you maximize for some other capability in the hardware. You wouldn't, for example, do a lot of still frames, because still frames (excised from motion footage) have flutter and related problems. But we say, "You need still frames? OK. Even though this technology is not great for it, we will use it." Because it will let us realize our imaginations.

What we are aiming for is a design that is not too constrained by the elements of the medium. This is a very hard thing to do, because when you work on a project, you want to do the best you can. So you start maximizing locally. And we don't want that. We don't want to make the best oil painting, we want to make the best painting, independent of what the substance is.

MP: How narrowly or widely circumscribed is your area of play? The company, as many know through press reports and marketing changes that have gone on lately, has de-emphasized education.

KH: I'm not sure that we have. I think what we've been doing is an analysis—thinking about how to organize to get all our businesses done. Education is still a major focus. It always will be.

MP: Is that because the base of the company is, or was, the IIe, the II GS, etc?

KH: The Apple II and now the MacIntosh (not the MacII) are strictly for education. What's interesting in my view is that experimenting in the education area lets you predict what business is going to be like. Because—where are all the knowledge workers now? Most of them are not sitting behind desks in offices. Most of the knowledge workers, per unit, are in schools. So it's the best place for us to work, now.
   If I can learn how to communicate information about the Constitution, I should be able to communicate personnel manuals, or whatever. That's somewhat straight-forward. It's the same set of techniques that will be used. So, education has wonderful usefulness about it. It's not a great business. The consumer market is the biggest market, if you really do your projections. The business market is certainly bigger, in terms of a possible applied base. Education is much smaller. There's a limited number of schools, and they have low budgets.
   But, to learn how to do these things, each of the markets has distinct advantages in terms of learning to think about the problems. We try to take advantage of all of them.
   Our lab is marketplace-independent, but we choose education examples very often because we know a lot about education, and also, I think, it's generalizable.

MP: That's very interesting. It's also an inversion of one of the things McLuhan prophesied. He said that the first place to apply a new technology would be business. And that was, of course, borne out in the early days of videodisc. In 1982, 1983 and 1984, when the arcade market crashed, at least IBM and AT&T were there doing training discs—followed later by Apple. After business, government gets involved. And finally, finally, schools adopt the new technology. This is a funny kind of flip you are suggesting.

KH: That's right. You take dollar revenue—it may come in the order you just described. It was government first in the videodisc area, in my experience, then it went to business, and finally to schools. The schools are barely in the videodisc business. But in terms of technique and way of doing things, which you can then sell across your markets, I think the education area is real core.

MP: Are you folks working on anything besides interactive videodisc?

KH: We're playing with CD-ROM. As a publishing medium, it's a wonderful way just to get data from here to there. And we're starting to use it as an access method. But so far videodisc is helping us quite a bit.

MP: Is that because it's relatively easy to do? You don't have to send the discs out for pressing. You can do that here.

KH: You are exactly right. We specialize in something we call Casual Multimedia Production. "Casual"—I love that. The phrase we use is: "We want to have a very fluid medium." We don't want a frozen medium, and that's the problem with optical media, they're frozen!

MP: Quite true.

KH: Libraries are frozen too. What I want to do is always have the database source material. Then, the real work is the manipulating, the changing and the moving. And, that's what makes us different from a movie company. We really want everyone to be making these things. We don't want just a couple of big artists. We want to have things people can make themselves.

MP: Philosophically, I am totally in agreement.

KH: Some people disagree completely and some people agree completely. (Laughter)

MP: It's a funny thing. The impetus that drives people to interactive (philosophically, my own politics—which dragged me into this craziness 20 years ago) is the empowerment of the viewer and the desire to turn television inside out, to create another species of experience. And yet, for several years in this particular sub-branch of interactive video—videodisc—I have found myself working with this extremely nonplastic, extremely immutable and invisible form. You don't know what a hard copy is until you get back a videodisc from a pressing plant!

KH: That's right. I know. It's crazy. So, this casual multimedia production that we are specializing in—offers low-end tools to make the product and change it. We discovered recently that none of us has the experience to know if we like using these things. But we know we love making them.

I have to make sure that we keep working from the materials. You don't do all the work on paper. Sometimes you do it on paper, and sometimes you put images onto a write-once videodisc right away. You mock 'em up on *HyperCard* right away, and you find out what the experience is like. You don't invest umpteen thousands of dollars before you know what the product is going to be. If you're doing Human Experience Design, which is what we're trying to do, you have to get it going very quickly. And then you have to—this can be very frustrating—do it through iteration. You don't do it by divine vision.

Maybe in a few years, some of us will have great divine vision, and we can work differently. Now, we're trying to use tools that let you see what the product looks like, that lets you change it and try it again, and so on.

MP: I believe that the OMDR is the way of the present.

KH: OMDR is wonderful, even if you never deliver a product on it. Having OMDR as a tool in an environment can be crucial. We talk about information labs for schools, for example, hoping that they have the same kind of equipment we have. And OMDR only costs about $20,000 now. That's a big number for a school but it's much less than the original price of $300,000. What it all means is that people can put things together differently. It's truly great.

MP: What you say is so exciting, because you are offering new angles to the meaning behind all this technology. New ways of thinking about it.

KH: Maybe we spend too much time thinking about it. We're flipping things a little bit, giving them a new go.

MP: This is a different perspective than the one I was able to deduce from the things I've seen (at Apple). You are actually up to something different.

KH: For better or for worse ...

MP: I think for the better.
What are you doing with video? I've seen the DNA demo, *Life Story*. I've seen the almanac. I've seen, LaserDisc's animal encyclopedia, in Tokyo, and I know you're playing with something similar. Is there anything innovative that you're doing with video itself?

KH: Probably not. As you saw in *Life Story*, even being able to put controls over the video, using the overlay in that way, is a stretch. The AST board lets you start putting the video into a window. There are a lot of interface issues that we're starting to play with. But, (to answer the question:) not really.

MP: That's not part of the mission. The mission is to explore the new dimension.

KH: That's right.

MP: Talk to me globally about linkages between computers and discs. Clearly, the next thing Apple's going to do technically is put it all on one screen. That will make life easier both in the business world and in the schools. But conceptually, what do you see going on?

KH: Again, the distinctions should blur. One shouldn't be thinking about it that way (i.e., in terms of video versus computers). I think one thing is real funny: you put controls on a video screen and you have one feeling. You put the images inside a computer screen and you feel different. It ought to be the case that it shouldn't matter.

It's just different elements that are available for design.

Now, on the one-screen, two-screen issue, I like two-screen systems lots of times. Just in terms of simultaneous comparison. I mean, I want a whole wall. The bigger I can get it, the better. So I think having one-screen systems will be great in many applications, and in those environments where you don't have room for the two. (Indicating the usual stack: laser disc player with monitor atop, and computer keyboard, floppy disk drives and monitor piled up at the side.) Though in fact, this is pretty tidy. It fits on a desk almost as well as a single screen.

MP: Control programs. What others are you using besides *HyperCard*?

KH: We're using *Course of Action*, and *VideoWorks*.

MP: Your focus is much broader than the educational one I knew of.

KH: What we try to do is make real things. The disadvantage is that when a project doesn't have the philosophy behind it, it just looks like itself. Each of our projects has been an experiment for us. *Life Story* is an experiment in how you take linear material and add some depth to it. We have others that are databases. The *Visual Almanac*, for example, is a database project. We basically say, instead of putting the movie on and seeing how you add interactivity and a database, let's put it (the movie) together with a database and see how you can add tools to make these things tell stories. So each of our projects is an experiment.

The almanac is a very interesting experiment right now because what we intend to do is get it out into the world. We have all the images cleared and all the tools working. And, we will just give it away, to let people play with it, because I think we're going to learn a lot from what other people do with it.

I have this list of the things that movies bring and the things that computers bring. It has terms like linearity, multiplicity of views, emotion and cognition, images and text, and so on—through all the different distinctions. And what is clear, after I have laid out all these dichotomies, is that there is good and bad in both columns. Singularity of view. Democracy in action. But in my view, we should have both. I ought not, as a designer, constrain myself to either of these. This is a caricatured way of talking about this, but we want to be able to say: this information is available for me. I can make presentations. We call those things "titles"—*Life Story* is a title, it's a person's point of view on how to teach about DNA. And then we have what we call techniques—sets of tools that let you take those things apart and put them together in new kinds of ways. So you end up with things that are a little movie-like, and a little computer-like. From each of those areas you pull whatever techniques you need.

Another thing that's relevant here is Negroponte's old three rings. You know, computing, publishing and entertainment, in overlapping circles. These things are not going to come together by themselves. That's the difference between an academic view and a business view. My argument is that here's the present; they're overlapping a little bit. Eventually they'll all be overlapping and that will be the information business.

Apple is over here (she indicates computing). Let's get ourselves in the center.

Let's get ourselves using all those areas. We're already working a lot in print. Let's push over here a bit (toward publishing), not just to get into that corner, but to get nearer the center. So you have a technology business here, the production of titles and materials here, and you have a whole publishing area. We do all of them.

So in the worst case, we're an over-worked center. Then what we talk about is computer-centered environments. You take this notion that you're sitting there in the middle as a computer, particularly if you're a computer company—that's our view—and you start adding things: you add scanners, CD-ROMs, laser discs, video monitors for output, audio sources—you start to build an environment. You add cameras as input. You did the same thing with desktop publishing, took a scanner, a computer and what we call Apple Cart, which is what is driving our CD-ROM and videodisc attempts. You come out on monitors and videotape recorders.

What you really want is a range of input devices for sights and sounds, and you want a range of output devices as well. The computer becomes a controlling element, it becomes where the human experience occurs.

MP: So, sum up if you will, from a producer's perspective. Where are we going?

KH: We're going to get tools that let us try things out quicker. The tools are going to be cheaper. We're actually going to get an installed base out there, so we can make a business out of this.

With higher volumes of production, our price per unit is going to come down, which is going to help on accessibility issues. Our display technology is going to surprise us. We are going to end up with very high resolution displays, and very big ones. HDTV is just a beginning to that. Five or ten years out, you start having the possibility of fiber optic links getting you the source imagery you need. We hope that the imagery will start to become accessible, so you won't have to be such a large production house to do this work.

Already with things like 8mm you start to see the right tools becoming available. To me, the quality of audio on 8mm is so nice. In the computer area, computer tools like *VideoWorks*, *Course of Action* or *HyperCard* start to become quite accessible to people for doing audio-visual management and presentation. Three years ago, before *HyperCard*, it wasn't so easy to control your images. You had to have a programmer, and all these different layers. Now one person can know a lot about the video production, the design and the programming. It creates people who are capable of following the whole project through. It takes away a lot of segmentation. So you don't end up with a computer culture in one part of your project, a video culture, education culture and corporate training culture in other parts. You end up with individuals who don't have to be whiz-bang programmers to be able to put things together. It's a real advantage.

The result allows you to work in small teams. And we're finding that a small team of three or four is about what you want. Of course, when you get down the production road, you need a lot more. It's the design phase that's being changed.

In the delivery phase, I think the displays will be better, whereas in the design phase you can start to have more flexibility and fluidity.

The flip side, incidentally, is we're going to start to see a couple of great examples.

Just like there are great movies that I can allude to, or in architecture, I can talk about certain great structures. We'll get enough good multimedia products completed, so I can describe to you what my project is without showing it to you. I can't (currently) describe what my project is without showing it to you. I can say, "Oh, now imagine here a picture of so-and-so, and if you click here, you get more information." It just doesn't work. Whereas, if I show it to you, you say, "Oh, I get it." It's simple.

As we get a common set of references, we'll start to be better at communicating. Hopefully, then, the people who are in the training business, or whatever, won't have such a big gap between what they can deliver and what they think they want to do. I think it's promising.

MP: The key is visualization tools.

KH: That's right. None of us have ever experienced these things, so how can we build them?

MP: Even the people who think they know what they are doing have trouble, all the way through the project, seeing what they are doing. So, this change in the design side is going to have a lot of impact.

KH: My sense is you also start to have some low budget productions that can be more experimental. This changes the culture, so you have a different kind of people involved. You don't have to mount a $300,000 project to do something interesting.

MP: Thank you, Kristina.

11 Agenda for the Future

We inhabit a world shaped by computers. But computers have not truly found their home in this world. Various explanations are offered:

- Computers must become more human, not we more computer-like.
- Computers are still hard to program.
- Computers are no more than costly typewriters and list-keepers, so why should I get one?

In truth, computing's failure to invade the hearth and heart of 80 million American homes must be sought at a deeper level—not everyone gets a kick out of making baked sand do their bidding.

With the NeXT machine, I can begin to speak to my box, and it can talk and sing to me. But I am primarily a visual hunter of information, as are most of my human brethren. I consume infinities of visual data before breakfast. And I insist that my computer meet me on this chosen sensory ground. Dance for me, smart-box. Pattern your intelligence so I can ingest with a glance everything you know. Computer-video can make this promise real. But how do we get from here to there? How do we grow from endless lists to a visual language—a tongue that reaches beyond the limits of spoken syntax?

Computer-video multimedia delivery systems are slowly coming along: CD-I bids to be an abler platform than first appeared. Intel is busily creating the chips to make DVI real and affordable. Capable authoring tools are coming out. The first commercial DVI applications have appeared. By 1993, there will be two million (plus) DVI-capable PCs out there, hungry for software. The first profits in interactive computer-based multimedia will then begin to be made.

151

So, again, what do we pathetic earthlings do while Moses (Intel) is up on the Mountain (IBM)? Worship at the golden calf of CD-ROM directory mania? Author ever-vaster multimedia almanacs? How many still pictures and B-level stereo audio segments can we cram onto a CD-ROM, with fractal compression? And who cares? These are the wrong questions, and the wrong goals. What we need to consider is what we lack to do visual computing:

- multimedia authoring systems,
- visual tools,
- multimedia operating systems,
- powerful applications, and
- visual indexing software.

At the 4th Microsoft CD-ROM Conference, one felt an embarrassing lack of guidance regarding the future. People came to that CD-ROM gathering hungry for hints and directions. The conference adopted the title "Seeing Is Believing" and was wrapped in the imagery of magicians at work. This was appropriate since what was purveyed as vision was in fact *presque-vue*—vague intimations of the nearly seen, but not yet clearly grasped. The enthusiasm of the sponsors, Bill Gates' in particular, brought thousands to the trough, but what was there to drink?

The most interesting software input was a list of proposed topics developed in a focus group by some high school students. Some of what they would love to see dealt with by a visual computer included:

- space exploration,
- career guidance,
- vacation guides and atlases,
- art instruction, and
- a human anatomy guide.

Beyond this list, there were no compelling demos offered and the visions of the young were not yet reflected in the dreams of the industry's captains.

At that 4th CD-ROM Conference I wondered: if the CD-ROM revolution is truly Bill Gates' obsession, then why will he not lead it? This industry needs more than an annual shot of info-amphetamines. It needs tools and product samples—not design examples. And most of all, it needs vision. As as result, I made some notes on needed developments that still ring true as follows:

Start with some righteous applications. Multimedia almanacs are cute, but they won't change, or even kindle, anything.

New ways of visualizing abstract ideas are needed, and new strategies for visual representation of multi-variate, dynamic processes are required. These can be basic tools in the kit of video PC users in the 1990s. One way to move toward this might be to take on the *Index of Knowledge* of the *Encyclopedia Britannica*, and to discover (i.e., invent) visual methods for describing the maternity and linkage of the major branches of human knowledge. This is a four-dimensional mapping problem. If it is solved correctly, it will offer a Rosetta Stone for the development of new intellectual indexing systems and new methods for representing (and inter-linking) abstract concepts.

All those who think "key words" are the way to handle the bottomless visual data resources of the mid-1990s, raise your hands (and close your eyes). Good luck.

After tinkering up some tools, the developers of multimedia's future could move on to visual authoring systems. In a stabler silicon environment, the innovators could develop languages for software authors—visual meta-languages for creating and controlling the ideographic utterances of our 1995 PCs.

Finally, for now, someone needs to invent the visual operating system that will

Interactive Video is the Hidden Dimension of the Idiot Box. When its Users Become the Controllers and Creators of Their Experience, TV will Fulfill its Destiny. Multimedia Computing and Videodisc Have Made This Transformation Inevitable, Technologically and Economically.

supplant DOS. Our multimedia maseratis won't long suffer a horse-and-buggy operating system. Big bites of visual patterned intelligence must be swiftly ingested, manipulated, synthesized and output by the able devices of the mid-1990s. In this connection, Microsoft has two choices: invent the future, or license it from someone else.

Henry Kissinger once wrote, "History is the story of people who sought their future in their past." McLuhan warned against the "rearview mirror" approach to the future, a characteristic posture of literary man. It is time to transcend the literary tradition. The future has come. "The age of macroscopic gesticulation is upon us," McLuhan taught.

Microsoft can lead the ascent, for about two percent of their gross revenues per year, over the next three to four years. Or, they can cede the high ground, and suffer the humiliation of having their early vision rented back to them. The future will be shaped by companies with vision. And they may well keep one final long list on CD-ROM—of those who hung on to the past when the plates of technology, and opportunity, slipped.

The 5th CD-ROM gathering came and went, with only modest progress to report on platforms and tools. No one invented the visual future. Intel did make good on certain board-level price and delivery promises regarding DVI. CD-I hit the exhibit floor in force with a dozen titles. Recordable desktop CD-ROM devices appeared. But Toshi Doi, inventor of the CD warned, "Almost no one knows how to make multimedia. We must set standards ... we have less than five years to make CD multimedia real, or all our work on CD will have been in vain."

Platforms are emerging. Tools are multiplying, too. The authoring systems we need to do CD-based multimedia are starting to arrive. And videodisc remains a viable base, technically and financially, to gain experience in producing interactive multimedia. Our mission is and will remain to go forth and be productive.

FINDING OUR WAY IN THE COMMUNICATIONS MEDIUM OF TOMORROW

Interactive video is more than an innovative wrinkle in television. It offers us new ways to think, to teach and to communicate. It has its own grammar, and we'd be wise at this early stage to admit we know very little about how to use it.

What little we know, a fraction of it codified above, would probably best be used defensively, not prescriptively. Think of this book as a map of a mine field. Use it to avoid blowing your next project. The language of interactive video is yours to invent. Look inward, and beyond what little we know, to shape a new species of communication.

Of course, there is no reason to ignore what's known. But interactive video is not ready for templates. It isn't even ready for theories. The words of another wise man ring in my head. The neurophysiological healer, Moshe Feldenkrais would constantly adjure his students: "Organize yourselves..." That is, use your knowledge to shape your intention and your every move. Because, he would add, "When you know what you are doing, you can do what you want." Only when you understand how and why you make the choices you do, can you act freely and creatively. If you operate by rote—if you misdirect the flashlight of knowledge, pointing it into your own eyes—you're a sleepwalker, driving by sensory rearview mirror into an unseen future.

"The real tomorrow of high-definition, wrap-around, user-controlled, multi-sensory virtual reality will belong to those who organize themselves appropriately and who dare stare into the bright sun, of the not-yet-known. Reaching for that future, let's be guided by the words of Hanns Kornell, a champagne maker in northern California:

> "To be in this business, you've got to love what you're doing, you've got to be half nuts, and you've got to put a little of yourself into every bottle."

Amen.

Part 4
Appendixes

Appendix A:
Glossary of Interactive Terms

The glossary of interactive terms presented here was originally developed by Jim Griffith of Media Learning Systems. It was later updated by Rockley Miller and published by *The Videodisc Monitor*. The version which follows has been further enlarged and updated for inclusion in this book.

ACRONYMS AND ABBREVIATIONS

ADPCM: adaptive differential pulse code modulation.
AI: artificial intelligence.
AIM: American Interactive Media.
AIV: advanced interactive video.
ANSI: American National Standards Institute.
ASCII: American Standard Code for Information Interchange.
AVC: Association of Visual Communicators.
A/D: analog-to-digital.
BER: bit error rate.
BIOS: basic input-output system.
BPS: bits per second.
CAA: computer augmented acceleration.
CAD/CAM: computer-aided design and computer-aided manufacture.
CADD: computer-aided design and drafting. *See CAD/CAM.*
CAI: computer-aided (or assisted) instruction. *See CBT.*
CAL: computer-aided learning. *See CBT.*
CAM: computer-aided manufacture. *See CAD/CAM.*
CAV: constant angular velocity.
CBI: computer-based instruction. *See CBT.*
CBL: computer-based learning. *See CBT.*

159

CBT: computer-based training.

CD: compact disc.

CD-3: 3-inch compact disc. *See compact disc.*

CD-A: compact disc-audio. *See compact disc.*

CD-DA: compact disc-digital audio. *See compact disc.*

CD+G: compact disc plus graphics.

CD-I: compact disc-interactive.

CD-IV: compact disc-interactive video.

CD-PROM: compact disc-programmable read only memory. *See compact disc-write once/read many.*

CD-ROM: compact disc-read only memory.

CD-WO: compact disc-write once.

CD-WORM: compact disc-write once/read many times. *See compact disc-write once.*

CDV: compact disc-video or CD Video.

CD Video LD: industry convention for 8-inch and 12-inch laser videodiscs with digital sound.

CED: capacitance electronic disc.

CGA: color graphics adapter. One IBM Personal Computer graphics standard.

CLUT: color look-up table.

CLV: constant linear velocity.

CP/M: control program/microcomputer.

CPU: central processing unit.

CRT: cathode ray tube.

CVD: compact videodisc.

D/A: digital-to-analog.

DAC: digital-to-analog converter.

DAT: digital audiotape.

dB: decibel.

DDP: Digital Data Processor (As in Sony's SFA/DDP) standard.

DOS: disk operating system.

DRAW: direct read after write.

DVI: digital video interactive.

EDAC: error detection and correction.

EEPROM: electronically erasable programmable read-only memory.

EGA: Enhanced Graphics Adaptor. An IBM Personal Computer graphics.

EIA: Electronics Industries Association.

EIAJ: Electronics Industries Association of Japan.

EIDS: Electronic Information Delivery System (US Army).

EPROM: erasable programmable read-only memory.

fps: frames per second.

HBI: horizontal blanking interval.

HDTV: high-definition television.

Hz: hertz.

IBM PC: IBM Personal Computer.

IBM PS/2: IBM Personal System/2.

ICVD: interactive compact videodisc. *See compact videodisc.*

ICW: interactive courseware (US Army).

ID: identification.

IEEE: Institute of Electrical and Electronics Engineers.

IGC: The Institute for Graphic Communications.

IIA: Information Industry Association.

IICS: International Interactive Communications Society.

ISD: instructional systems design.

ISO: International Standards Organization.

ITVA: International Television Association.

IVIA: Interactive Video Industry Association.

IV: interactive video.

IVD: interactive videodisc.

IVT: interactive videotape.

K or KB: kilobyte.

LAN: local area network.

LD: laser disc.

LDP: laser disc player.

LSI: large-scale integration.

LV: LaserVision. *See reflective optical videodisc.*

LV-ROM: laser vision-read only memory.

LASER: light amplification by stimulation of emission of radiation.

MHz: megahertz.

MIA: Multimedia Industries Association

MS-DOS: Microsoft Disk Operating System. *See disk operating system.*

NAB: National Association of Broadcasters.

NAPLPS: North American Presentation Level Protocol Standard.

NTSC: National Television Systems Committee (of the EIA).

ODDD: optical digital data disc.

OEM: original equipment manufacturer.

OMDR: optical memory disc recorder (Panasonic).

OROM: optical read-only memory.

OS: operating system. *See disk operating system.*

PAL: phase alternation line.

PC-DOS: (IBM) Personal Computer Disk Operating System. *See disk operating system.*

PC: personal computer. *See also IBM Personal Computer.*

PMMA: polymethyl methacrylate.

POI: point-of-information. *See point-of-purchase.*

POP: point-of-purchase.

POS: point-of-sale. *See point-of-purchase.*

PROM: programmable read-only memory.

PS/2: *See IBM Personal System/2.*

RAM: random-access memory.

RGB: red-green-blue.

ROM: read-only memory.

rpm: revolutions per minute.
RTOS: real-time operating system.
SALT: Society for Applied Learning Technology.
SCSI: small computer systems interface.
SECAM: "sequential couleur a memoire" (sequential color with memory).
SEG: special effects generator.
SFA: still-frame audio. *See compressed audio.*
SMPTE: The Society of Motion Picture and Television Engineers.
S/N: signal-to-noise.
STI: speech transmission index.
TED: TelDec electronic disc.
VBI: vertical blanking interval.
VCR: videocassette recorder.
VGA: Video Graphics Array. IBM PS/2 graphics standard.
VHD: video high density.
VHS: video home system.
VITC: vertical interval time code.
VLSI: very large scale integration.
VTR: videotape recorder.
WORM: write-once/read-many.
WYSIWYG: "what you see is what you get."

GLOSSARY OF INTERACTIVE TERMS

access: To retrieve information from a storage medium, such as videodisc, computer disk or tape, or videotape. *See also random access.*

access time: The total time required to find, retrieve and display data after initiation of a retrieval command. Access time is usually measured at its worst, or the longest possible time it takes to get from one section of the medium (tape, disc, disk) to another. This is generally a matter of minutes on videotape, two or fewer seconds on videodisc or CD, and milliseconds or microseconds on a computer. *See also random access.*

active program: The length of audio and video program material on the master videotape not to exceed the one-side capacity of a videodisc. For reflective optical videodiscs, the maximum active program length is about 30 minutes for CAV, and 60 minutes for CLV. For transmissive optical videodiscs, the maximum active program length is approximately 27 minutes.

active video lines: All video lines not occurring in the horizontal and vertical blanking intervals.

A/D: Abbreviation of analog-to-digital. The conversion of data or signal storage from one format or method to another.

address: (1) Usually an alphanumeric or numeric label that identifies a location where information is stored. (2) A time code or frame number that identifies the location of video and/or audio material on tape or disc.

address code: (1) Time code that indicates each video frame by hour, minutes, seconds and frame number. (2) The picture, chapter or still cue code inserted in the vertical blanking interval of the videodisc frame and read by the disc player.

advanced interactive video (AIV): Interactive videodisc format and system using LV-ROM, a method of storing analog video, digital audio and digital data on a single videodisc. The system was developed by Philips UK, the British Broadcasting Corporation, Acorn Computer and Logica Ltd. Most prominent application was the BBC's Domesday Project.

algorithm: A plan of the exact sequence of steps needed to accomplish any task.

algorithmic: Type of precisely defined or structured procedure (usually mathematic) which provides the solution to a problem in a finite number of steps. *See also heuristic.*

aliasing: (1) Undesirable visual effects (sometimes called artifacts) in computer-generated images, caused by inadequate sampling techniques. The most common effect is jagged edges (jaggies) along diagonal or curved object boundaries. (2) The stair-step effect on a raster display system which lacks the resolution to reproduce diagonals or circles as smooth images.

alphanumeric: Characters which include the letters of the alphabet, numerals and other symbols used for codes, punctuation and mathematical operations.

analog: The representation of numerical values by physical variables such as voltage, current, etc. Information which steadily flows and changes. Continuously variable quantities whose values are analogous to the quantitative magnitude of the variables. Analog devices are characterized by dials and sliding mechanisms. Contrast with digital.

analog video: A video signal that represents an infinite number of smooth gradations between given video levels. By contrast, a digital video signal assigns a finite set of levels. *See also video, digital video.*

anti-aliasing: Software adjustment to make diagonal or curved lines appear smooth and continuous in computer-generated images.

application: The use of a technology to accomplish a defined purpose. Common applications for interactive video are in training and marketing.

applications program or applications software: A computer program designed to do one specific job (i.e., accounting system, training program, etc.).

architecture: In computing, a term which refers to the design or the design philosophy of computer hardware. In personal computing, "open architecture" indicates systems (such as the IBM PC and compatibles) which allow for the addition of peripherals and internal enhancement cards from third-party hardware vendors. "Closed architecture" systems (such as the original Apple Macintosh computer) are touted as "all-in-one-box" solutions, and users are discouraged from adding memory, additional ports, etc. In some cases, even opening the machine for routine maintenance will void manufacturer's warranty.

archival: Readable (and sometimes writeable) for an extended period. Archival media have defined minimum life-spans over which the information will remain stable (i.e., accurate without degradation).

artificial intelligence (AI): (1) A computer software programming approach which allows a machine to use accumulated experience and information to improve its own operations. (2) The development or capability of a machine that can proceed or perform functions which are normally associated with human intelligence, such as learning, adapting, reasoning, self-correction, automatic improvement.

artwork: The still illustrations or graphics prepared for printed work, film, or video. Includes sketches, drawings, captions, titles, photos, maps, graphs, charts, etc. Sometimes termed "flat art."

ASCII: American Standard Code for Information Interchange; the standardized, eight-bit data character code system used internationally to code alphabetic, numerical and other symbols into the binary values used in computer applications.

aspect ratio: The measurement of a film or television viewing area in terms of relative height and width values. The aspect ratio or most modern motion pictures varies between 3 x 5 to as large as 3 x 7, which creates a problem when a wide-format motion picture is transferred to the more square-shaped television screen, with its aspect ratio of 3 x 4. HDTV has an aspect ratio of roughly 3 x 5.

audio: Pertaining to sound, the audio portion of a television program, the audio track, an audio source tape.

audio track: The section of a videodisc or tape which contains the sound signal that accompanies the video signal. Systems with two separate audio tracks (most videodiscs) can offer either stereo sound or two independent soundtracks.

author: (1) To prepare a computer program, often with an authoring language or authoring system. Such systems or languages allow users without formal computer programming training to prepare applications programs for computer or videodisc-based systems. Authoring is a structured approach to developing all elements of an interactive video or videodisc program, with emphasis on pre-production. (2) The individual who creates an interactive videodisc control program.

authoring language: A specialized, high-level, plain English computer language which permits non-programmers to perform the programming function of courseware development. The program logic and program content are combined. Generally provides fewer capabilities or options than an authoring system.

authoring system: Specialized computer software which helps its users design interactive courseware in everyday language, without the painstaking detail of computer programming. In an authoring system, the instructional logic and instructional content are separate. Allows greater flexibility in courseware design than an authoring language.

auto-repeat: A switch-selectable feature of many videodisc players that allows automatic and continuous replay of discs until interrupted.

auto-start: A feature of many videodisc players that automatically begins program play when the disc is loaded into the player.

auto-stop: A pre-programmed instruction that tells the videodisc player to stop automatically on a designated frame. Also known as picture stop.

bandwidth: The range of signal frequencies that a piece of audio or video equipment can encode or decode; the difference between the limiting frequencies of a continuous frequency band. Video uses higher frequency than audio, thus requires a wider bandwidth.

bar code: A block of optically coded parallel lines, read by a scanner or wand which transmits a coded message to a microprocessor. Bar codes are most familiar in retail sales, but also appear in workbooks, labels and other printed matter.

BASIC: Beginner's All-purpose Symbolic Instruction Code, an algebraic computer programming code developed at Dartmouth College. BASIC is a conversational type of programming language that uses English-like statements and mathematical notation.

baud: Commonly used unit of transmission speed to describe the rate at which binary data is communicated. One baud is approximately equal to one bit per second. Common baud rates are 300, 1,200 and 2,400 bps (bits per second).

Betacam: A half-inch video recording format developed by Sony that offers near one-inch tape quality on a portable system.

binary code: A code in which every element has only one of two possible values, which may be the presence or absence of a pulse, a one or a zero, or a high or a low condition for a voltage or current.

bit: Contraction of BInary digiT. A unit (either 0 or 1) of information equal to one binary decision. The smallest unit in computer information handling. It can be a single character in a binary number, a single pulse in a coded group of pulses, or a unit of information capacity. A computer's processing capability is usually determined by the number of bits which can be handled at one time. Personal computers, for instance, commonly offer 8-, 16- or 32-bit microprocessors.

bit density: The number of bits of digital data that can occupy a given volume or area of storage medium.

blanking: A period where no video signal is received by the monitor, while the videodisc player searches for the next video segment or frame to display.

board: An internal plug-in unit for printed-circuit wiring and components. Computer boards often control either some essential function of the computer's central processor or provide a special feature such as a modem or a videodisc player interface. Also known as card. *See also interface.*

branch: (1) To jump from one sequence in a program to another. (2) A video segment selected on the basis of viewer response.

branching point: A location in a program where the viewer may select among two or more optional paths or destinations.

byte: A generic term developed by IBM to indicate a measurable number of consecutive binary digits which are usually operated upon as a unit. Bytes of eight bits ("by eight") usually represent either one character or two numerals. A computer's storage capacity or memory is figured in kilobytes (K or KB). A typical personal computer might have a 16-bit word length (two byte) central processor and 512KB memory.

capacitance: (1) Electrical capacity. (2) The ability to store an electric charge. Two incompatible videodisc formats (see CED format and VHD format) use variations in capacitance between the disc and the pickup stylus or sensor to transmit recorded video and audio data.

capacitance electronic disc: *See CED format.*

capacitor: The component of an electronic circuit which stores and releases voltage.

card: *See board.*

carrel: A semi-enclosed booth which functions as a study station with a desk-like work area, often containing interactive video hardware, a video monitor, an audio headset, a microcomputer, and reference or work books. Also known as work station.

cathode ray tube (CRT): A type of graphic display which produces an image by directing a beam of electrons to activate a phosphor-coated surface in a glass or plastic vacuum tube. The picture tube of any television or video monitor.

CAA: Computer Augmented Acceleration. An upgraded version of the CLV disc mode. CAA disc speed is held constant for vast majority of the time. All speed changes required for high-density recording are made in a series of short steps. Between each step, the number of video trace lines being recorded for each turn of the disc is held to a constant whole number. This feature prevents the horizontal sync information from being located in adjacent tracks where it could produce the "barber pole" (see cross-talk) effect. *See also CLV, CAV.*

CAI: Computer Assisted Instruction. Interactive learning without video. The use of the computer in the instructional process.

CAV: Constant Angular Velocity. A mode of videodisc playback in which a disc rotates at a constant speed, regardless of the position of the reading head or stylus. Thus, each frame is separately addressable. In optical videodisc technology, each track contains two video fields that comprise one complete video frame. CAV discs revolve continuously at 1,800 rpm (NTSC) or 1,500 rpm (PAL), one revolution per frame. Program time is 30 minutes per side on a 12-inch disc, 14 minutes per side on an 8-inch disc. All VHD discs run in CAV, at 900 rpm (NTSC) and 750 rpm (PAL). Each track contains four fields (i.e., two frames). All transmissive optical videodiscs are also formatted for CAV play. *See also CLV, CAA.*

CD-ROM: *See compact disc-read only memory.*

CD-ROM drive or CD-ROM player: A device that retrieves data from a disc pressed in the CD-ROM format. Differs from a standard audio compact disc player by the incorporation of additional error correction circuitry. Often lacks the necessary D/A converter to play music from standard compact discs.

CED format: The CED (capacitance electronic disc) system has grooved media and uses a stylus in physical contact with the disc surface, reading capacitance signals embedded on the disc. Developed for the consumer marketplace by RCA under the trade name SelectaVision; abandoned by RCA in 1984.

chapter: One independent, self-contained segment of a computer program or interactive video program.

chapter number: Numbers displayed on the screen which identify individual videodisc chapters.

chapter number code: A number encoded in the vertical blanking interval of the disc frame, which allows chapter numbers to be displayed on the screen during play.

chapter search: A function of most videodisc players which allows specific chapters to be accessed by chapter number.

chapter stop: A code imbedded in some videodiscs to signal the break between two separate chapters, to allow access to specific chapters.

check disc: A videodisc produced prior to quantity replication to confirm the accuracy of interactive program encoding.

chip: A crystalline silicon device in which microscopic electronic circuitry is printed photographically to create passive and active devices, circuit paths and device connections within the solid structure.

chrominance: The color information in a full-color electronic image. Requires luminance, or light intensity, to make it visible. *See also luminance.*

CLUT: Color look-up table. A selection of colors assigned a digital value and held in a table. The program decodes a color picture for display by matching the code stored for each pixel with the associated color value in the look-up table. This technique reduces the amount of information needed to recreate a color image.

CLV: Constant Linear Velocity, an alternative format for reflective optical videodiscs. CLV (or "extended-play") discs allow twice as much play time (up to one hour) per side, but many of the user-control capabilities of the CAV format are forfeited (e.g., no still-framing is possible). The CLV disc can be read in linear play only, but can provide chapter search capability. With CLV videodiscs, the revolution speed varies with the location of the pickup to ensure a constant data rate. CLV optical videodiscs vary in speed from 1800 rpm at the inner track to 600 rpm at the outer edge. Program time is 60 minutes per side (12-inch disc) or 20 minutes per side (8-inch disc). All compact discs are played in CLV mode. *See also CAV, CAA.*

combi-player or combination player: A single player that can accommodate a variety of disc formats. Units currently available can play 8-inch and 12-inch videodiscs as well as 4.75-inch compact discs and CD Videos.

compact disc (CD) or compact audio disc: A 4.75-inch (12 cm) optical disc that contains information (usually musical) encoded digitally in the CLV format. Popular format for high fidelity music, offering 90 dB signal/noise ratio, 74 minutes of digital sound, and no degradation of quality from playback. The standards for this format (developed by NV Philips and Sony Corporation) are known as the Red Book. The official (and rarely used) designation for the audio-only format is CD-DA (compact disc-digital audio). The simple audio format is also known as CD-A (compact disc-audio). A smaller (3-inch) version of the CD is known as CD-3.

compact disc + graphics (CD+G): A CD format which includes extended graphics capabilities as written into the original CD-ROM specifications. Includes limited video graphics encoded into the CD subcode area. Developed and marketed by Warner New Media.

compact disc-interactive (CD-I): A compact disc format which provides audio, digital data, still graphics, and limited motion video. The standards for this format (developed by NV Philips and Sony Corporation) are known as the Green Book.

compact disc-read-only memory (CD-ROM): A 4.75-inch laser-encoded optical memory storage medium with the same constant linear velocity (CLV) spiral format as compact audio discs and some videodiscs. CD-ROMs can hold about 550 megabytes of data. CD-ROMs require more error-correction information than the standard prerecorded compact audio disc. *See also OROM.* The standards for this format (developed by NV Philips and Sony Corporation) are known as the Yellow Book.

compact disc-video (CDV or CD Video): A CD format introduced in 1987 that combines 20 minutes of digital audio and six minutes of analog video on a standard 4.75-inch compact disc. Since its introduction, many firms have renamed 8-inch and 12-inch videodiscs CDV, capitalizing on the consumer popularity of the compact audio disc.

compact disc-write-once (CD-WO): Proposed. A variant on CD-ROM that can be written to once, and read many times. Under development by NV Philips and Sony Corporation. Also known CD-Write Once/Read Many (CD-WORM).

compact disc-write once/read many (CD-WORM): *See compact disc-write-once (CD-WO).*

compact videodisc (CVD): Under development. An analog/digital hybrid capable of delivering interactive mixed-media applications. Can deliver either 10 (CAV) or 20 (CLV) minutes of full motion video. Developed by SOCS Research, licensed by Mattel for toy and other applications. Originally known as interactive compact videodisc (ICVD).

compatible: Used to describe different hardware devices that can use software or play programs without modification.

composite video: The complete visual wave form of the color video signal composed of chromatic and luminance picture information; blanking pedestal; field, line and color sync pulses; and field equalizing pulses. As opposed to RGB display, a type of computer color display output signal comprised of separately controllable red, green and blue signals. *See also RGB.*

compress: To reduce certain parameters of a signal while preserving the basic information content. Compressing usually reduces a parameter such as amplitude or duration of the signal to improve overall transmission efficiency and to reduce cost.

compressed audio: A method of digitally encoding and decoding up to 40 seconds of voice-quality audio per individual disc frame, resulting in a potential for over 150 hours of audio per 12-inch videodisc. By using a buffer to store the audio information, a limited amount of audio may be delivered to accompany a still-frame image. Also known as still-frame audio.

compressed video: A video image or segment that has been digitally processed using a variety of computer algorithms and other techniques to reduce the amount of data required to accurately represent the content, and thus reducing the space required to store that content.

computer: A data processor that performs predefined computations and data manipulations (including arithmetic and logic) under the control of a program, usually without intervention by a human operator during a processing run.

computer-augmented acceleration: *See CAA.*

computer-based training (CBT): The use of a computer to deliver instruction or training. Also known as computer-aided (or assisted) instruction (CAI), computer-aided learning (CAL), computer-based instruction (CBI) and computer-based learning (CBL).

computer graphics: Visual images generated by a computer. Graphics standards for IBM-compatible personal computers include CGA, EGA, Hercules and VGA.

color map: A table which stores the definitions of the red, green and blue (RGB) components of colors in a computer graphics system to be displayed on the monitor.

conditional branching: An instruction which causes the video to branch if a specified condition or set of conditions is satisfied. If the condition is not satisfied, the computer proceeds in its normal sequence of control. A conditional transfer also tests the condition.

constant angular velocity: *See CAV.*

constant linear velocity: *See CLV.*

consumer market: Also known as the domestic market; that segment of the videodisc market in which home players and discs (films, entertainment, games) are used primarily in private homes. The other side of this coin is the professional, or industrial market.

continuous branching: A program that enables the user to modify the presentation at any point rather than at specific branch points.

continuous motion video: *See full-motion video.*

control track: A synchronizing signal on the edge of the videotape which provides a reference for tracking control and tape speed during playback.

courseware: Instructional software, including all discs, tapes, books, charts and computer programs necessary to deliver a complete instructional module or course.

central processing unit (CPU): The central processor or brain of a computer system, in which all calculations, instructions and data manipulations are performed. It contains the main storage, arithmetic unit and special register group. In a microcomputer, the CPU is often on a single chip called a microprocessor.

crawl: Alphanumeric text that moves across a TV screen, either horizontally or vertically. The term is generally used to mean a steady controlled text movement, such as the display of credits at the end of a program.

cross talk: The pickup of adjacent track data by the laser head as it attempts to hold the desired signal track being played. In laser videodiscs, cross talk appears as a mesh of transparent, wormy lines that drift across the picture. Caused by inaccurate focusing of the laser beam on the disc surface, from the warping of discs, dirt, etc.

cue: A pulse entered onto one of the lines of the vertical blanking interval that results in frame numbers, picture codes, chapter codes, closed captions, white flags, etc. on the master tape and/or videodisc.

cue inserter: The device which places cues on lines of the vertical blanking interval of the master tape to tell the disc-mastering equipment in which field it should put the frame ID code on the disc.

cues, chapter: A set of nine pulses placed in the vertical blanking interval of the master videotape which identify a tape frame as the first frame of a new chapter. When the disc-mastering equipment reads a chapter cue, it encodes a chapter stop command on the vertical interval of the corresponding disc frame. Chapter cues cannot be used independently of picture cues.

cues, picture: The first set of nine pulses in the vertical blanking interval of the master tape which identify the start of a complete frame. Each time one of these picture cues passes through the mastering equipment, it triggers a frame counter, and the disc-mastering equipment encodes the frame number on the disc's vertical interval.

cues, still: A third set of nine pulses in the vertical blanking interval of the master videotape which identify the coming frame and tells the disc-mastering equipment to place a code on the disc to automatically switch the disc player to freeze-frame mode. Like the chapter cue, it cannot be used independently of the picture cue. Chapter cues and still cues may be used independently of each other.

cursor: The image on a computer screen which indicates where information may be entered next or the user's position on the screen. Generally takes the form of a solid or flashing rectangle, but may be a different icon as well.

cybernetics: The comparative study of control and communication in information-handling machines and the nervous systems of animals.

D/A (digital-to-analog): The conversion of data or signal storage from digital form to analog.

data: A general term used to denote any or all facts, numbers, letters and symbols that refer to or describe an image, object, idea, condition, situation or other factors. It connotes basic elements of information which can be processed or produced by a computer.

database: A collection of organized data that can be accessed by computer.

data channel: The bi-directional data path between the input/output devices and the main memory in a digital computer that permits one or more input/output operations to transpire concurrently with computation.

data rate: The speed at which data is transmitted.

debug: To locate and correct any errors, or "bugs" in a computer program, or to detect and correct malfunctions in the computer itself.

decibel (dB): A logarithmic measure of the ratio between two powers, voltages, currents, sound intensities, etc. Signal-to-noise ratios are expressed in decibels.

decision tree: A system based on the premise that decisions spawn outcomes which require other decisions. Thus, a single choice feeds into a network of other decisions.

delivery system: The set of video and computer equipment actually used to deliver an interactive video program. A delivery system may comprise as little as a videodisc player with (or without) on-board microprocessor, a monitor and a keypad—or may extend to an external computer, two or more monitors, and a variety of peripherals.

density: The closeness of space or data distribution on a storage medium, e.g., a magnetic drum, magnetic tape, optical disc, or CRT. The higher the density, the higher the resolution—or more compact the storage.

desktop: Refers to the miniaturization, simplification, and/or personalization of a previously large, complex device or task. The microcomputer is known as a "desktop computer," since it was the first computer that literally could sit on top of a desk. Systems which combine sophisticated layout software with laser printer output are known as "desktop publishing" systems. Apple's *HyperCard* can make simple computer/videodisc authoring available to the unskilled user, and has been called "desktop authoring."

device driver: Software that tells the computer how to talk to a peripheral device, such as a videodisc player or printer.

dialog or dialogue: An exchange of information between human operator and the program/technology. The term is mostly used to imply a continuing interactive response between both operator and program/technology.

digital: A method of representing signals by a set of discrete numerical values, as opposed to a continuously fluctuating current or voltage. *See also analog, sampling.*

digital audio: Audio tones represented by machine-readable binary numbers rather than analog recording techniques. Analog audio is converted to digital using sampling techniques, whereby a "snapshot" is taken of the audio signal, its amplitude is measured and described numerically, and the resulting number is stored. More frequent sampling results in a more accurate digital representation of the signal.

digital-to-analog converter (DAC): Interface to convert digital data represented in discrete, discontinuous form to analog data represented in continuous form.

digital video: A video signal represented by computer-readable binary numbers that describes a finite set of colors and luminance levels. *See analog video, video.*

digital video interactive (DVI): A technology that allows compression and decompression of graphics and full-motion video for recording on CD-ROM, videodiscs, magnetic disks, or other digital storage media. Developed by RCA's David Sarnoff Research Center. Currently owned by Intel and featured on chips and circuit boards for IBM PCs.

digitize: To register a visual image or real object in a format that can be processed by a computer; to convert analog data to digital data. *See scan.*

disc: Flat, circular rotating medium that can store and replay various types of information, both analog and digital. Disc is often used in reference to optical storage media, while disk refers to magnetic storage media. Often used as short form for videodisc or compact audio disc (CD). *See also videodisc, compact disc and disk.*

disc directory: A picture frame at the beginning of a videodisc which identifies the program content by chapter and/or frame numbers.

disc geography: The determination of the relative placement on the disc of motion and still frame material and how the viewer is routed through it. Closer placement of related segments results in faster access time between them.

disk: Magnetic storage device, as in hard disk or computer diskette.

disk operating system (DOS): A computer operating system designed for use with a disk. Languages, applications and utility programs can be transferred quickly between CPU memory and the disk storage. DOS is the preferred usage for both the Microsoft disk operating system (MS-DOS) and IBM's personal computer disk operating system (PC-DOS), both designed for the IBM personal computer (and essentially the same language).

display: A device (usually a CRT) upon which numbers, characters, graphics or other data are presented.

documentation: The written records and/or instruction manual for computer or videodisc programs.

DRAW (direct read after write): Type of optical discs that can be locally recorded but not erased. Recorders use a high-power laser to burn pits into a heat-sensitive layer beneath the surface of a recordable disc. Information is then read by a lower-power laser.

drive: That part of a computer-based system (such as a personal computer) into which floppy disks, tapes, CD-ROMs, videodiscs, other optical or magnetic media are inserted when they are being used to input, process or output information. *See also hard disk.*

drop frame: A system of modifying the frame counting sequence to allow the time code to match real time clock. Not used in IVD production.

dual-channel audio: The capability to reproduce two audio channels, playing them either simultaneously or independently. A characteristic of all optical videodisc systems.

dub: (1) The copying and combining of visual and audio elements to produce a composite, properly balanced (mixed) master tape. (2) The copying of a tape to produce a new generation.

dump: (1) The process by which digital program code is transferred from a videodisc to a Level 2 videodisc player's microprocessor. It is usually read from the second video channel. (2) A unit of program code that is loaded into the videodisc player's microprocessor at one time. A single videodisc may contain multiple program dumps. *See Levels of Interactive Systems—Level 2.* (3) To remove; to stop; to empty.

DYUV or delta-YUV: An efficient color-coding scheme for natural pictures used in CD-I. The human eye is less sensitive to color variations than to intensity variations, so DYUV encodes luminance (Y) information at full bandwidth and chrominance (UV) information at half bandwidth or less, storing only the differences (deltas) between each value and the one following it.

edit: To link one piece of audiotape or videotape to another, or to create a master tape of a video program, usually from a variety of source media.

Electronic Information Delivery System (EIDS): A combination microcomputer/videodisc-based audio-visual training system, contracted by the U.S. Army in November 1986 as the Army-designated standard stand-alone computer-based instructional system. EIDS refers both to the hardware configuration, provided by Matrox Electronics under the initial contract, and to the standard for courseware (i.e., EIDS-compatible).

electronic publishing: The delivery of information via computer or other electronic medium; closely parallels the concept of traditional print publishing, except for medium of delivery. Content is held in a storage device for delivery on a computer screen, rather than printed on paper.

emulator: A system used to test interactive videodisc programs prior to mastering. It uses videotape (or other storage devices such as write-once discs) under computer control to simulate the final disc operation.

enabling objectives: A required, subsidiary objective that contributes to the achievement of a terminal objective.

encode: (1) To convert information to machine or computer-readable format (frequently binary numbers) representing individual characters or groups of characters in a message. Encoding is one step in the process of converting an analog signal into a digital signal. The three steps are sampling, quantizing and encoding. (2) To combine three color signals into one composite video signal.

encryption: A procedure for encoding data that makes it difficult to decode data without proprietary software or hardware. This procedure protects data or software from unauthorized access or use.

feedback: (1) The reinforcement of correct responses or the correction of errors by the videodisc system or instructor. (2) Audio squeal resulting from placement of a source microphone in the proximity of an output speaker.

fiber optics: The technology in which audio-visual signal/information is transmitted through a glass fiber strand capable of transmitting light.

field: One-half of a complete television scanning cycle (1/60 of a second NTSC; 1/50 of a second PAL/SECAM); i.e., all of the odd or all of the even scan lines of a raster image. When their scan lines interlace, two fields combine to make one video frame.

field dominance: In videodisc pre-mastering, the order of the video fields established on the videotape during edits or transfers. A tape with field one dominance has a new picture beginning on field one; with field two dominance, the new picture begins on field two. The field dominance of the master tape determines on which field the videodisc frames will begin. *See also flicker and interfield.*

field/frame synchronization: The elimination of video and film frame ambiguity by the use of full-frame identification process during film-to-tape transfer.

field frequency: The rate at which a complete field is scanned or displayed, normally 59.94 times a second in NTSC.

film chain: A term used to encompass the total grouping of equipment used to transfer slide or movie film picture frames to electronic picture frames; usually consists of film and slide projectors, a multiplexer and a television camera. Also known as telecine or telecine projector.

flicker: Video effect (usually unwanted) on a still or frozen frame caused when the two fields that combine to make one video picture frame are not identically matched, thus creating two different pictures alternating every 1/60 of a second. Interfield flicker can occur when field dominance is incorrectly specified or if field dominance changes at one or more points on the master tape, after being edited on equipment that is incapable of frame-accurate editing. Also known as jitter. *See also interfield, interfield jitter.*

flow chart: A diagram or map of interactive logic which represents the possible paths a user may take through a program; comprised of standard symbols for program segments, decision points, responses and logic flow.

font: In printing and computer-based character generators, a particular style set of alphanumeric characters and symbols (e.g., Times Roman, Helvetica). Many devices support a variety of font options.

frame: A single, complete picture in a video or film recording. A video frame consists of 525 lines (NTSC) or 625 lines (PAL/SECAM), running at 30 frames per second (NTSC) or 25 fps (PAL/SECAM). Film runs at 24 fps.

frame address: In videotape and optical videodiscs, each frame has an address or frame number. A frame address is put on each disc or tape frame in the form of a frame address code. On videodisc, these are sequential integers from 1 to 54,000. On tape, the SMPTE frame address in in the form 01:00:00:01 (hour:minute:second:frame).

frame address code: A code located in the vertical blanking interval of a video frame.

frame buffer: (1) An apparatus capable of storing all 525 lines of a television frame. (2) A memory device which stores, pixel by pixel, the contents of an image. Frame buffers are used to refresh a raster image. Sometimes they incorporate local processing ability. The "depth" of the frame buffer is the number of bits per pixel, which determines the number of colors or intensities which can be displayed. Frame buffers can also serve as time base correctors.

frame rate: The speed at which video frames are scanned or displayed; 30 frames a second for NTSC; 25 frames a second for PAL/SECAM.

freeze-frame: A single frame from a segment of motion video or film footage held motionless on the screen. Unlike a still frame, a freeze-frame is not a picture intended to appear motionless, but is one frame taken from a longer motion sequence.

full-frame time code or non-drop frame time code: A standardized SMPTE method of address-coding a videotape that retains all frame numbers in a chronological order, resulting in a slight deviation from clock time.

generation: (1) In hardware, a stratification referring to major evolutionary stages in specific products or product areas. Successive generations incorporate significant improvements upon the original (first-generation) or successive models. (2) In storage media (tape, disc, etc.), the number of times a reproduction is removed from the original source, the first generation being made from the original recording. The second is a copy made from the first-generation copy, and so on. With analog information, each generation usually results in a degradation of image or sound quality. With digital formats, there is no difference between generations.

generic courseware: Educational courses that are not specific to one organization and thus appeal to a broader market; as opposed to custom courseware, which primarily meets the needs of one specific client or audience.

generic videodisc: Videodisc material that can be used with courseware developed by more than one organization; discs associated with the subject matter but not with a particular course.

giga-: Prefix meaning "one billion," as in gigabyte (one billion bytes).

graphics: All still visuals and animated motion sequences prepared for a production. Includes camera cards, slides, electronically generated letters and symbols, and special graphical set pieces.

graphics input: Data used to create or alter a graphics display; entered through such interaction devices as keyboard, data tablet, button device, touch screen or light pen.

graphics input device: Unit such as a digitizer which gives the computer the points which make up an image so that it may be stored, reconstructed, displayed or manipulated.

graphics output device: Device used to display or record an image. A display screen is an output device for "soft copy"; hard copy output devices produce paper, film or transparencies of the image.

hardware: The electric, electronic and mechanical equipment used for processing data. The complement of hardware is software.

high-definition television (HDTV): Any one of a variety of video formats offering greater visual accuracy (or resolution) than current NTSC, PAL or SECAM broadcast standards. Current formats generally range in resolution from 655 scanning lines to 2,125 scanning lines, having an aspect ratio of 5 : 3 (or 1.67 : 1), and a video bandwidth of 30 MHz to 50 MHz (5+ times greater than NTSC standard). Digital HDTV has a bandwidth of 300+ MHz. HDTV is subjectively comparable to 35mm film.

hertz (Hz): The standard unit of measuring frequency. One Hz is equal to one cycle (or vibration) per second. One kilohertz (KHz) equals 1,000 cycles per second, and one megahertz (MHz) equals 1,000,000 cycles per second. Named after German physicist Heinrich Hertz (1857 to 1894).

heuristic: Pertaining to methods of trial-and-error style problem-solving by evaluation of the progress made toward the final result. *See also algorithmic.*

high-level language: A computer programming language designed to facilitate use of computers by people. One statement in a high-level language may be translated to many assembly language or machine code instructions. High-level languages are also designed to be machine independent, in contrast to assembly level language. Examples of high-level languages are BASIC, Pascal, FORTRAN, etc.

HyperCard: A Macintosh-based software product developed by Apple Computer. Using the philosophy of hypertext, the program enables users to randomly organize information in "stacks."

hypermedia: An extension of hypertext that incorporates a variety of other media, in addition to simple text.

hypertext: The concept of non-sequential writing which allows writers to link information together through a variety of paths or connections. Hypertext allows users to seek greater depths of information by moving between related documents along thematic lines or accessing definitions and bibliographic references without losing the context of the original inquiry. The term was coined by Theodore Nelson in the early 1960s. Hypertext is the driving concept behind Apple's *HyperCard* software program.

IBM Personal Computer (PC): A commercially available microprocessor-based computer, the most popular business microcomputer available through 1989. IBM has discontinued sales of its PC series in favor of its new Personal System/2, PS/2 featuring

MicroChannel architecture. However, several other hardware vendors still support the popular PC standard.

IBM Personal System/2 (PS/2): The line of IBM personal computers introduced in 1987. The Personal System/2, which uses either a 80286, 80386 or 80486 processor depending on model, replaces the IBM Personal Computer series. The PS/2 was designed to run MS-DOS and to be compatible with existing PC software. Higher level PS/2 models will use OS/2, the new multi-tasking operating system designed by IBM and Microsoft.

icon: A symbolic, pictorial representation of any function or task.

IEEE-488: General purpose interface bus; eight-bit parallel interface between host computer and graphics terminal.

individualized instruction: Software that modifies its instructional method or content based on student feedback, to optimize learning.

industrial market: That segment of the videodisc market that includes non-home applications (i.e., training, education, sales, corporate communications, etc.). Also known as the professional market.

information processing: The processing of data representing information, and the determination of the meaning of the processed data.

information retrieval: The ability to choose interactively any data segment, and to have a computer find it instantly. This can be provided via a table of contents, or menu, at the beginning of each segment or by an alphabetically ordered index.

in-house: Actions performed entirely within a given company or organization, using its own resources, facilities, and expertise.

insert edit: Type of edit in which new video and/or audio material is inserted into any point of a pre-existing material recorded on the master tape. No new time code or control track is recorded.

instant jump: The feature of some videodisc players that allows branching at imperceptible speeds between frames within certain minimum distances, usually one to 200 frames away. The branch occurs during the vertical blanking interval between images.

instructional designer: Overseer and developer of any educational videodisc- or computer-based program.

instructional design: The methodology and approach used to deliver information in a manner that achieves learning objectives. Aspects include question strategy, level of interaction, remediation strategy, reinforcement and branching complexity.

intelligent videodisc player: A videodisc player with processing power and memory capability built into it. *See also Levels of Interactive Systems.*

interactive: Involving the active participation of the user to direct the flow of a video or computer program; a system which exchanges information with the viewer, processing the viewer's input in order to generate an appropriate response within the context of the program; as opposed to linear.

interactive media: (1) Media which involves the viewer as a source of input to determine the content and duration of a message, permitting individualized programing. (2) A philosophy of media production designed to take maximum advantage of random access, computer-controlled videotape and videodisc players.

interactive video: (1) Video that is under user control—delivered by videodisc, tape, cable, or CD. (2) The fusion of video and computer technology. A video program and a computer program running in tandem under the control of the user. In interactive video, the user's actions, choices and decisions genuinely affect the way in which the program unfolds.

interactivity: A reciprocal dialog between the user and the system; in theory, simultaneously resulting in mutual transformation.

interface: (1) The link between two pieces of disparate equipment, usually the computer's central processor and its peripherals. (2) The device or circuit that provides the communicating transition between different systems.

interfield frames: A product of the 3 : 2 pull-down, film-to-tape transfer process, in which the video frame is composed of two fields each of a different film frame. These mixed fields do not interfere with normal viewing, but on a videodisc, where a viewer can freeze on any single frame, an interfield frame might produce unwanted flicker.

interfield jitter: Rapid (60 times per second) successive display of fields from two different frames.

interlace: The pattern described by the two separate field scans when they join to form a complete video frame. As the video picture is transmitted, the first field picks up even-numbered scan lines—the second, odd-numbered ones. The two lace together to form a single, complete frame.

interleaving: A method of storing information sequences in an alternating series of frames and playing the sequences using instant jump capabilities to achieve continuous play of a cohesive segment. This procedure allows instant branching between different linear segments.

jaggies: *See aliasing.*

jitter: *See flicker.*

joy stick: Graphics input device that positions the cursor or initiates a program transformation by means of a control lever. Popularly used in video and arcade games.

jump: *See branch.*

jukebox: A machine that can hold and play several items (such as vinyl records, videodiscs, and compact discs).

karaoke: Japanese for "empty orchestra;" sometimes called "music minus one." A form of entertainment where guests in a karaoke bar or home may take up the microphone and sing. The videodisc not only provides background music but offers mood enhancing visuals and lyrics as well. Karaoke has been a major hit in Japan and a mainstay of the Japanese videodisc business since the introduction of the videodisc in the early 1980s.

keyboard: A panel containing alphanumeric and other keys used to create text and convey instructions to a computer. The computer keyboard usually contains, in addition to the familiar characters and symbols of a typewriter keyboard, keys dedicated to specific computing functions.

keyer: Signal processing device which cuts a hole in a video picture and fills in the hole from a different video source, e.g., computer-generated text and graphics keyed over NTSC video.

keypad: A small keyboard or section of a keyboard containing a smaller number of keys.

kilobyte (K or KB): A term indicating 1,024 bytes of data storage capacity. A typical personal computer might have 640K of memory. *See byte, bit.*

kiosk: A small, stand-alone retail vending station, usually manned by one individual who dispenses items from a limited inventory of products (newspapers, magazines, candy, cigarettes). The term is used in interactive video applications to describe the housing for an unmanned, self-contained, free-standing interactive system that is generally located in a public access area.

landing pad: A range of frames within which a player can locate a frame or frame sequence, accessed from other parts of the disc.

laser: (1) Light Amplification by Stimulation of Emission of Radiation. An amplifier and generator of coherent energy in the optical, or light, region of the spectrum. In the laser videodisc system, a laser is used to read the micropits on the videodisc which contain the picture and sound information. (2) Generic name for reflective optical videodisc format promoted by NV Philips, Pioneer LaserDisc, Sony, and others. *See optical videodisc, reflective optical videodisc; also videodisc formats.*

laser disc (LD): Common name for reflective optical videodisc. LaserDisc is a trademark of Pioneer Electronics USA for its reflective optical videodisc products.

laser rot: The degradation of a laser (video- or compact) disc due to either improper process control or raw material contamination, or both. Phenomenon of early videodiscs. Manufacturers now seem to have the problem under control.

LaserVision (LV): Trade name for reflective optical videodisc format promoted by NV Philips, Sony, Pioneer, Hitachi, and others.

LaserVision-read only memory (LV-ROM): Format (developed by Philips for the BBC Domesday Project) which combines analog video and digital data and audio on 12-inch optical reflective videodiscs. *See advanced interactive video (AIV).*

learning station: A physical location such as a study carrel, which contains special materials and equipment used by a student to learn.

levels of interactive systems: Three degrees of videodisc system interactivity proposed by the Nebraska Videodisc Design/Production Group in 1980.

Level 1 system: Usually a consumer-model videodisc player under keypad control with still/freeze frame, picture stop, chapter stop, frame and chapter address, and dual-channel audio, but with limited memory and limited processing power.

Level 2 system: An industrial-model videodisc player with the capabilities of Level One, plus on-board programmable memory and improved access time.

Level 3 system: Level 1 or 2 players interfaced to an external computer and/or other peripheral processing devices. (Note: some commentators have advocated additional levels—4, 5 and up—suggesting that the addition of digital audio, touch screens, etc. creates new levels of interactivity. However, the industry has not settled on any single standard for these higher levels, and any innovation mentioned with such "higher levels" all fall categorically into Level 3.)

levels of interactivity: Derived from levels of interactive systems. Levels 1, 2 and 3 refer to the interactive design features available with each respective hardware configuration. Does not relate to the quality, relative value or degree of sophistication.

light pen: Stylus with a light-detecting mechanism used as a graphics input or user interface device to identify cursor position on a CRT.

Line 21 technology: Line 21 of the vertical blanking interval, used for captions, full-page text information, Infodata caption system; can provide second language information.

linear: A motion sequence designed to be played from beginning to end without stops or branching, like a film; as opposed to interactive or user-controlled.

local area network (LAN): A system which connects two or more microcomputers to allow shared resources and communication.

location: (1) In computing, the place where data can be recorded or found, usually discussed in terms of address. (2) In video and film production, a place where material is shot or filmed in an environment that represents the "real" world, as opposed to the studio.

logic: The basic principles and applications of truth tables, the relationships of propositions, the connection of on-off circuit elements, etc., for mathematical computation in a computer.

loop: The repeated execution of a series of instructions for a fixed number of times—avoid recursive loop: an inescapable bind.

low-level language: A programming language in which each statement is translated into a single machine instruction; also called an assembly language. *See also language.*

luminance: Brightness; one of the three image characteristics coded in composite television (represented by the letter Y). May be measured in lux or foot-candles.

machine language: A set of binary codes used to express computer instructions and data in a directly executable form. No further translation to a lower level language is required to execute.

magnetic tape: A thin, strong, non-elastic tape coated with a ferromagnetic emulsion, which can record, store and play back information of various kinds. Audiotape records sound. Videotape holds sound and pictures, as well as electrical signals used in editing and in interactive video applications. Can also be used as a computer data storage device.

magneto-optics: An information storage medium that is magnetically sensitive only at high temperatures, while stable at normal temperatures. A laser is used to heat a small spot on the medium, allowing a normal magnet to change its polarity. The ability to tightly focus the laser greatly increases the data density over standard magnetic media.

main menu: A table of contents or index listing major modules of information available on a videodisc. Usually done with a still frame.

main trunk: A program's principal course or line of direction.

mainframe computer: Originally, the main framework of a computer's central processing unit. Subsequently, the CPU itself. Popularly, the largest of computing devices, both in size, capacity and cost. *See also minicomputer, microcomputer.*

master: (1) An original audiotape, videotape or film. Used for broadcast or to make copies. (2) The process of producing master, mother, and stamper videodiscs, which are used for replicating videodiscs.

master videodisc: First stage disc in videodisc manufacture. On the master disc, conductivity to receive the converted video signal is produced by evaporation or plating. The disc is then nickel-plated.

medium: In computing, a substance or object on which information is stored. Its plural is media.

mega-: Prefix meaning "one million," as in megabyte (one million bytes).

megahertz (MHz): A million hertz (million cycles per second).

memory: The location in which computer-based equipment stores recorded information, either permanently or temporarily. Usually measured in kilobytes or megabytes.

menu: A table of contents or index listing major modules of information available on a videodisc or computer program.

microcomputer: A small computer containing a microprocessor, input and display devices, and memory all in one box. It may or may not be connected to a host computer and/or peripheral devices. Sometimes referred to as a desktop, personal or home computer. However, microcomputers have also found huge acceptance in the business community.

microsecond: One-millionth of a second.

millisecond: One-thousandth of a second.

minicomputer: Usually, a minicomputer is a parallel binary system with smaller computer storage, slower processing speeds, and lower cost than large mainframe computer systems, but larger in each of these aspects than microcomputers.

modeling: An educational process whereby a computer-based learning system is used to represent another system or process. The learner can change values and observe the effects of the change on the operation of the system.

modem: Contraction of MOdulator/DEModulator. Digital device that converts data from a computer into signals that can be transmitted over ordinary telephone lines, and vice-versa.

modular: Consisting of independent units that are part of a total structure.

monitor: A CRT or RGB screen which may accept either video signals, computer display information, or both.

mother disc: Second stage disc in videodisc manufacture. From the nickel-plated master, a mother disc is formed and coated with remover. The mother disc is then nickel-plated. From the mother, several daughters are made and used in pressing.

motivational device: A design element which arouses and sustains interest or regulates activity for the purpose of causing the student to perform in a desired way.

mouse: A hand-held, rolling remote control device for a computer which guides the cursor on the computer screen.

MS-DOS/PC-DOS: The disk operating systems of IBM Personal Computers, developed by Microsoft Corporation. *See disk operating system.*

multiscan (or multisync) monitor: A video display that accepts a wide variety of horizontal and vertical timings, from NTSC video to RGB computer graphics. Multiscan monitors will often automatically adjust to the appropriate timing.

nanosecond: One-billionth of a second.

NAPLPS: North American Presentation Level Protocol Standard (pronounced "nap-lips"). Visual computer standard for graphic communication protocol that provides a method for creating pictures and compressing them into relatively short blocks of digital data for storage and transmission over low-bandwidth channels. NAPLPS is one standard protocol for videotex.

noise: Random spurts of electrical energy or interference. In video, noise may produce a random salt-and-pepper pattern over the picture. Heavy video noise is called snow.

nondrop frame time code: SMPTE standardized method of address-coding a videotape by numbering all frames in a chronological order; this results in a slight deviation from clocktime. *See full-frame time code.*

NTSC: National Television Systems Committee of the Electronics Industries Association (EIA) that prepared the standard specifications approved by the Federal Communications Commission, in December 1953, for commercial color broadcasting.

NTSC format: A color television format having 525 scan lines; a field frequency of 60 Hz.; a broadcast bandwidth of 4 MHz.; line frequency of 15.75 kHz.; frame frequency of 1/30 of a second; and a color subcarrier frequency of 3.58 MHz. *See also PAL, SECAM.*

OEM: (1) Items which are purchased from the "original equipment manufacturer" (usually at substantial discount under retail) and often resold or repackaged under a different brand name. Does not refer to the manufacturer, but to the system integrator that resells the device. (2) To integrate and resell another firm's hardware.

offline: (1) Any operation not under the control of a computer, occurring independently. (2) Editing a workprint of original master footage without the use of a computer assisted edit system.

OMDR: Acronym for the Matsushita/Panasonic line of Optical Memory Disc Recorders. An 8-inch or 12-inch write-once videodisc which is not compatible with laser videodisc players from Pioneer, Sony Hitachi and Philips.

online: Equipment, devices, and systems in direct interactive communication with a computer.

optical digital data disc: A catch-all phrase for any optical disc used to store digital information.

optical disc player: The playback device for an optical videodisc.

optical memory: A generic term for technology that deals with information storage devices that use light (usually laser-based) to record, read, or decode data.

optical videodisc: A videodisc that uses a laser light beam to read information from the surface of the disc. The information in optical videodiscs is encoded in the form of microscopic pits pressed into the disc surface. The pits or holes modulate the laser in a manner that can be decoded by the videodisc player. Information stored in these pits is "read" by a laser beam and transmitted to a decoder in the player. *See reflective optical videodisc, transmissive optical videodisc.*

OROM or optical read-only memory: A 5.25-inch laser-encoded optical memory storage medium, which features a concentric circular format and constant angular velocity (CAV). OROMs have faster access time than CD-ROM discs, but less storage space (250 megabytes as opposed to 500). *See also CD-ROM.*

OS-9: A "real-time" operating system on which the CD-I operating system is based.

overlay: A term used to describe the keying of computer-generated text and/or graphics onto motion or still video.

package: A set of compatible, interlinked equipment designed to make up a complete delivery system, or a set of computer programs needed to handle one specific job, such as videodisc to computer interface, or a software application.

PAL format: Phase Alternation Line; the European standard color system, except for France. *See also NTSC, SECAM.*

palette: In digital video, the total number of colors available for pictorial presentations.

parallel: Simultaneous computer data processing of more than one part of a whole.

Pascal: A computer programming language designed to help teach programming as a systematic discipline and to do systems programming. Based on the language ALGOL, it emphasizes aspects of structured programming.

PC-compatible: Most often refers to computers compatible with the IBM Personal Computer standard.

performance objective: A tightly defined goal with a very low possibility of the learner misunderstanding what is to be done.

peripheral unit or peripheral: Equipment controlled by the computer, but physically independent of it (i.e., keyboard, printer).

personal computer: An inexpensive, somewhat portable computer for business and home use made popular by Apple and IBM. Also known as PC. *See microcomputer.*

photodiode: A device used in an industrial standard laser videodisc player to translate variations in the light reflected off the pitted surface of the disc into the electronic signals which comprise the audio, video and control tracks of the program.

picture stop: A function of some videodisc systems which allows the player to stop automatically on a specific frame during play; an instruction encoded in the vertical blanking interval on the disc to stop on a predetermined frame.

pit: The microscopic physical indentation or hole found in the information layer of a videodisc. They are cut in a spiral pattern on the inside surface of an aluminized plastic disc. (10 pits = 1 bit.) Pits on reflective optical discs modulate the reflected beam. Pits in transmissive discs block the beam or allow it to pass through the disc. Pits on VHD discs cause a detectable change in electrical capacitance. In all cases, variations in the pits carry the information.

pixels: An abbreviation of picture element. The minimum raster display element, represented as a point with a specified color or intensity level. One way to measure picture resolution is by the number of pixels used to create images.

PMMA: Polymethyl methacrylate. A rigid, transparent acrylic plastic used to manufacture many laser videodiscs.

point-of-purchase (POP): Interactive video units set up in public areas to demonstrate products, sell goods, advertise service, offer information, etc. to passers-by. Sometimes called point-of-sale (POS).

POP: *See point-of-purchase.*

port: The socket at which cables connecting the computer and its peripherals are attached.

portability: Refers to the ability of video courseware to be used on a variety of hardware systems. The greater the portability, the lesser the need for program modification to accommodate different hardware.

post-production: The stage in the preparation of a film or video program after the original footage has been shot. Can include editing, encoding, computer program authoring, etc.

pre-mastering: The stage in the production of a videodisc when the master tape (generally 1-inch Type C NTSC master helical videotape) is checked and prepared for final transfer onto the master disc, from which all subsequent discs will be pressed.

pre-production: All design tasks (flowcharting, story-boarding, scriptwriting, software design, etc.) that lead up to the actual shooting of material on video or film.

production: In video terms, the period when video or film footage is actually shot. *See also pre-production, post-production.*

program: (1) To plan a computation or process for computer operations, including coding, numerical analysis, specification of printing formats, etc. (2) A set of instructions or steps that tells the computer exactly how to handle a complete problem. (3) Material on a tape or disc viewed by an audience.

program map: For planning an interactive videodisc, a program map specifies which audience sees which topics/materials.

public domain: The condition of being free of copyright or patent. Public domain software is free of charge; anyone who wants to may use it.

quadruplex or quad: A videotape system, developed by Apex, which uses four video heads mounted 90° apart on a drum which spins at 240 revolutions per second (NTSC) or 250 rps (PAL). Quad uses broadcast quality 2-inch videotape.

quantize: A step in the process of converting an analog signal into a digital signal. This step measures a sample to determine a representative numerical value that is then encoded. The three steps are sampling, quantizing, and encoding.

random access: The ability to reach any piece of data on a storage medium in a very short time.

random-access memory (RAM): That part of a computer's memory that can both read (find and display) and write (record) information, and which can be updated or amended by the user; the largest part of a computer's memory, used in its day-to-day work. *See also read-only memory.*

raster: The area illuminated by a scanning beam of a TV system; grid; a raster display device stores and displays data as horizontal rows of uniform grid or picture cells (pixels). *See also vector graphics.*

read-only memory (ROM): (1) A computer storage medium which allows the user to recall and use information (read) but not record or amend it (write). (2) The smaller part of a computer's memory, in which essential operating information is recorded in a form which can be recalled and used (read) but not amended or recorded (written). *See also random-access memory.*

real estate: In interactive video technology, the space available (frame capacity) on a videodisc.

real time: The actual time in which a program or event takes place. In computing, refers to an operating mode under which data is received and processed and the results returned so quickly that the process appears instantaneous to the user.

reflective optical videodisc format: Reflective laser videodiscs contain their information imbedded as pits or holes in reflective surfaces sandwiched between layers of polymethyl methacrylate (PMMA). The shiny surface reflects the laser light into a mirror, which in turn reflects it to a decoder. The clear PMMA protects the information from dirt and superficial scratches. Reflective discs must be turned over to read information on both sides. Because this format uses a laser instead of a stylus to retrieve information, there is no physical contact between the reading mechanism and the disc itself, hence no wear or degradation during playback. Promoted by NV Philips, Sony, Pioneer, Hitachi, and others. *See also Laser, LaserVision, optical videodisc, videodisc formats.*

remediation: Corrective teaching.

remote control: Command of a computer or interactive videodisc program through an electronic device independent of the computer or disc player (i.e., keypad, touch screen, joy stick, mouse).

replicates: Videodisc copies pressed from the stamper disc.

replication: The mass reproduction of videodiscs or compact discs.

repurposing: The process of modifying the content of an existing program to accomplish a task other than the one for which it was originally designed. Often, Level 1 consumer videodiscs are repurposed for Level 3 use. Sometimes, linear materials are repurposed for IVD delivery in training, sales exhibit or games settings.

resolution: Number of pixels per unit of area. A display with a finer grid contains more pixels and thus has a higher resolution, capable of reproducing more detail in an image.

response-paced: Computer or videodisc-based instruction that teaches the student at a speed matching his/her demonstrated ability to learn.

RGB (red-green-blue): A type of computer color display output signal comprised of separately controllable red, green and blue signals; as opposed to composite video, in which signals are combined prior to output. RGB monitors typically offer higher resolution than composite. *See also composite video.*

RS-232C: A standard serial interface between a computer and its peripherals. Various peripherals (including some videodisc players) equipped with an RS-232C computer port can be plugged directly into a compatible computer. The RS-232C is becoming a standard feature of computers and their peripherals.

RTOS: Real-Time Operating System. CD-I operating system developed by Microware.

safe area: That area in the center of a video frame which is sure to be displayed on all types of receivers and monitors. Televisions and other monitors made at different times and by different companies are slightly different in size and shape, and the outer edge of the video frame (about 10 percent of the total picture) is not reproduced in the same way on all sets. Scanning voltages vary: it's a 60 year old technology, so there's variance in picture height and width.

sampling rate: (1) Rate at which slices are taken from analog signals when converting to digital. (2) Frequency at which points are recorded in digitizing an image. Sampling errors can cause aliasing effects.

saturated colors: Strong, bright colors (particularly reds and oranges) which do not reproduce well on video, but tend to bleed around the edges, producing a garish, unclear image.

scan: (1) In basic television and video transmission, the rapid journey of the scanning spot back and forth across the inside of the screen to form scan lines. (2) In interactive videodisc technology, a mode of play in which the player skips over several disc tracks at a time, displaying only a fraction of the frames it passes. Scanning can be done in forward or reverse. (3) In data capture, the process by which a document or hard copy image is converted to machine-storable image format. *See digitize.*

scan conversion: The process of putting data into grid format for display on a raster device. A two-ended picture tube reads in an image that is electrostatically stored, and then the other end scans the information out. This can now be performed with solid state CCD technology. This technique allows 525-line NTSC programs to be converted to PAL.

scan lines: The parallel lines sloping across the video screen from upper left to lower right, along which the scanning spot travels in picking up and laying down the video information which makes up the picture on the monitor. NTSC systems use 525 scan lines to a screen; PAL systems use 625 lines.

SCSI: Small computer systems interface (pronounced "scuzzy"). A device-independent interface used for a wide variety of computer peripherals.

search: In interactive video systems, to request a specific frame, identified by its unique sequential reference number, and then to instruct the player to move directly to that frame (forwards or backwards) from any other point on the same side of the disc.

search time: The amount of time required by a computer or disc player to locate specific data in the storage medium.

SECAM format: "Sequential couleur a memoire" (sequential color with memory), the French color TV system, also adopted in Russia. The basis of operation is the sequential recording of primary colors in alternate lines. *See also NTSC, PAL.*

segment: Any material with a start and stop frame; a motion sequence as well as a series of still frames meant to be accessed together.

SelectaVision: Trade name for RCA's defunct CED videodisc format. Originally used for their home videotape system.

sequence: An orderly progression of items of information; two or more frames forming a unit, e.g., motion sequence, still-frame sequence.

serial: The time-sequential handling of individual items.

shared disc or sharedisc: Videodisc which is produced jointly by several parties, each receiving a portion of the disc space for its own purpose. Often produced to bring down production costs.

shareware: Computer software sold with the intention that the user will pay a fee to the program author if the user is satisfied with the program. Also known as user-supported software.

signal-to-noise (S/N) ratio: The strength of video and/or audio signal in relation to the interference (noise) it has picked up as it passes through electrical circuitry. The higher the S/N ratio, the better the quality of the signal.

simulation: Representation of a system, sub-system, situation or device, with a degree of realism. The simulation mode enables users to learn the operation of equipment without damaging it or harming themselves or others. Extremely useful in training applications which involve potentially dangerous activities.

slow scan: Very slow rate analog or digital video system. The still picture is gradually painted onto the screen. Also known as freeze-frame video.

slow motion: A mode which allows the user to move forward or backwards through a video sequence at an exaggeratedly slow speed. In videodiscs, the player repeats a frame a specific number of times before automatically moving on to the next frame.

SMPTE time code: An 80-bit standardized edit time code adopted by SMPTE, the Society of Motion Picture and Television Engineers. *See time code.*

snow: *See noise.*

software: The programs, routines, subroutines, languages, procedures, videodiscs, charts, workbooks—in fact, everything that isn't hardware used in a computer or videodisc system.

speech recognition: *See voice recognition.*

speech synthesizer: A device that produces human speech sounds from input in another form.

stair-stepping: Jagged raster representation of diagonals or curves; corrected by anti-aliasing.

stamper: Disc made from mother disc which is used to mold final replicated discs. The stamper must be as flat as possible for reducing the warp and distortion of a disc.

stand-alone: Equipment such as a computer terminal or interactive video system (*see Levels of Interactive Systems—Level 2*) which is independent of any larger network.

step frame or step: (1) A function of optical videodisc players which permits the user to move either forward or reverse from one frame to the next. (2) To advance one frame forward or reverse.

still frame (1) A single film or video frame presented as a single, static image. (2) Refers to information recorded on a frame or track of a videodisc that is intended to be retrieved and displayed as a single, motionless image. Playback is achieved by repeating the play of the same track, rather than going on to the next; as opposed to a freeze-frame, which stops the action within a motion sequence.

sub menu: Allows viewer to branch to new information without returning to the main menu or detailed menu. Frequently used at the end of a module, located a good distance from the main menus or to offer the viewer the opportunity to review what has just been seen.

substrate: The molded plastic portion of a videodisc or compact disc.

surrogate travel: One application of interactive videodisc in which physical travel is simulated using disc and computer, allowing the user to control the path taken through an environment. Surrogate travel disc systems have been used for tours of nuclear power plants and for high-level security systems. Also known as vicarious travel.

tape, lead-in: In videodisc programs, the 40 to 60 seconds of video black preceding the active program.

tape, lead-out: In videodisc programs, a minimum of 30 seconds of video black with no audio following the active program.

TED or TelDec format: An early videodisc system developed jointly by Germany's Telefunken and Britain's Decca. It employed a flexible plastic foil disc read by a prow-shaped stylus. However, the first discs were only 10 minutes long, and picture quality was poor. The system appeared only briefly on the European market in 1975.

teleconferencing: A general term for meetings not held in person. Usually refers to a multi-party telephone call, set up by the phone company or private source, which enables more than two callers to participate in a conversation. The growing use of video allows participants at remote locations to see, hear and participate in proceedings, or share visual data.

teletext: Computer information inserted into the normal broadcast signal, usually during lines 18 to 21 of the vertical blanking interval. It is a one-way information system.

terminal: The point of communication between the user and a computer-based information system through which information can be input or output. In computing, terminals are often remote work stations connected to a central processing unit. A typical computer terminal includes a visual display unit, keyboard, and one or more disk drives.

terminal performance objective: The final, valued results the student seeks to achieve from the educative process.

test: Any strategy by which a response to the video courseware is elicited, and which results in the measurement of understanding by program logic.

3 : 2 pull-down: A method for overcoming the incompatibility of film and video frame rates when converting or transferring film (shot at 24 frames per second) to video (shot at 30 frames per second). The first film frame is actually exposed on three video fields, and the next film frame is exposed on two fields, the next film frame on three fields, the next on two fields, and so on. Thus, two of every five video frames will consist of fields that contain information from two different film frames. The resulting effect is noticeable as flicker during freeze frame use on videodisc.

time code: A frame-by-frame address code time reference recorded on the spare track of a videotape or inserted in the vertical blanking interval. It is an eight-digit number encoding time in hours, minutes, seconds and video frames.

touch screen: A video and/or computer display which acts as a control or input device under the physical finger touch of the user. Basic functions are executed by touching or stroking certain parts of the screen, and specific responses made by touching appropriate words, messages, or pictures as they appear. Different touch screen technologies use infrared grids, small wires separated by air spaces, changes in electronic capacitance, acceleration detection, plastic membranes, and other methods.

track: (1) A specific area of audiotape or videotape which contains program information or technical control information. (2) The accuracy with which a recorded videotape plays back.

transactional: Describes kiosks and other public machines at the retail level that can allow the user to purchase an item using cash or credit card.

transmissive optical videodisc format: Transparent videodisc which allows the laser beam to pass through the disc to the detector. Originally developed by Thomson/CSF (France), which introduced a player that could read both sides of a disc by changing the laser's focal point. Each side can contain 49,999 frames for 27 minutes of playing time, and an overall total of 99,998 frames are 54 minutes per disc. The only transmissive system to reach the market in the United States was the McDonnell Douglas Electronics Company (MDEC) LaserFilm system, which used one-sided discs. *See also optical videodisc, videodisc formats.*

transparency: Reduction of the user's perception of the computer system in the process of interaction. Machines which are extremely easy and inviting to use make the technology transparent.

turnkey: An off-the-shelf product or system that is ready to run when delivered—simply "turn the key."

user bits: Undefined bits within the 80-bit SMPTE time code word that are available for uses other than time coding.

user-friendly: Computer programs or systems which are designed for simple operation by non-technical users.

validation: The measurement and evaluation process by which courseware is refined before it is completed for distribution.

varied repetition: Design elements that repeat a segment of a lesson in a different manner to enhance learning.

VCR: Videocassette Recorder. Generic term for home videotape device.

vertical blanking interval (VBI): Lines 1 through 21 of video field one and lines 263 through 284 of field two, in which frame numbers, picture stops, chapter stops, white flags, closed captions, etc. may be encoded. These lines do not appear on the display screen, but maintain image stability and enhance image access. *See also horizontal blanking interval.*

vertical interval time code (VITC): SMPTE time code inserted in the vertical blanking interval between the two fields of a tape frame. This method eliminates errors that occur from tape stretch when using longitudinal time code.

vertical markets: Markets that demand specialized products suited to their professional needs. The medical and legal professions are two examples of vertical markets that have their own jargon and information needs. By contrast, office management products cater to a broader, more generalized market.

VHD format: Video High Density. A grooveless capacitance videodisc system which uses a broad stylus to pick up information. VHD discs rotate at a constant 900 rpm, contain four video fields per revolution, and can accommodate one hour of material per side without loss of special features. The discs are housed in a jacket which is inserted into the player and then removed, leaving the disc. The format, developed and marketed by Matsushita/JVC, is now available only in the Japanese market. The same player can handle both NTSC and PAL format discs. *See also capacitance.*

VHS or video home system: Consumer videotape format developed by JVC.

vicarious travel: *See surrogate travel.*

video: A system of recording and transmitting information which is primarily visual, by translating moving or still images into electrical signals. These signals can be broadcast (live or pre-recorded) using high-frequency carrier waves, or sent through cable on a closed circuit. The term video properly refers only to the picture—but as a generic term, usually embraces audio and other signals which are part of a complete program. Video now includes not only broadcast television, but many non-broadcast applications—such as corporate communications, marketing, home entertainment, games, teletext, security, and even the visual display units of computer-based technology.

videocassette: A compact playing/recording case which contains magnetic tape used in broadcast, industrial and home videotape applications. (Found in Beta, VHS, BetaCam, 8mm, Hi-8, 3/4-inch, D2 and 1-inch formats)

videodisc: A generic term describing a medium of information storage which uses thin circular plates of varying formats, upon which video, audio, and data signals may be encoded (usually along a spiral track) for playback on a video monitor.

videodisc formats: Reflective optical videodisc or laser; transmissive optical videodisc; CED or capacitance electronic disc; VHD or video high density (*see separate entries under each*).

video display: Television-type CRT (raster format) which decodes and displays information from a video source signal.

video head: The unit within a videotape player which reads video signals recorded on the tape.

videotape: Magnetic tape used to record video and audio.

videotex: (1) A collective name for systems which use the domestic TV receiver to display data from a central computer transmitted to the home set, either via coaxial cable or telephone link. (2) A special set of fairly low-resolution text and graphic characters that can be displayed via specific decoders.

virtual: Existing or resulting in effect though not in actual fact. In computing, a virtual device may reside only in memory while representing a hardware peripheral. The use of virtual devices can help programmers avoid hardware incompatibilities as actual configurations evolve.

virtual reality: Advanced simulation systems with computer generated 3D visual "worlds" presented to the user while deriving information about the user with motion and position sensing devices worn by the user.

voice-activated: Computer or videodisc program executed or controlled by the sound of a human voice.

voice recognition: A computer input technology in which a human utterance is recognized within the computer terminal and then converted into machine-usable binary code.

volatile storage: A storage device in which stored data is lost when operating power is intentionally or accidentally removed.

VTR: Videotape recorder.

white flag: A code inserted on videotape that identifies a new full frame, used when transferring film to videotape. Also known as full-frame ID.

window: (1) A segment into which the interactive videodisc user may enter at any point without missing the chapter stop. (2) A defined portion of a display screen in which a video image or other information may be shown.

work station: Display console with keyboard and input devices. *See also carrel.*

WORM (write-once/read-many) memory: A type of permanent optical storage that allows the user to record original information on a blank disc, but does not allow erasure or change of that information once it is recorded.

write: To transcribe recorded data from one place to another, or from one medium to another. Information from the computer is written to a disk, rather than on a disk.

WYSIWYG: "What You See Is What You Get" (pronounced "wizzy-wig"); refers to graphic display mode in some desktop publishing applications in which the page on the screen shows exactly how the printed page will appear.

X-Y coordinates: Points on a plane which has been divided across (X axis) and down (Y axis) on a pre-determined two-dimensional scale. X-Y coordinates can be used to plot a drawing or graph on a computer screen. A third axis (Z) is used in three-dimensional spatial plotting.

X-Y zoom: A digital system device giving the operator control over the scale of X and Y axes, thereby enlarging or reducing the image size.

YUV color system: A color encoding-scheme for natural pictures in which the luminance and chrominance are separate. The human eye is less sensitive to color variations than to intensity variations, so YUV allows the encoding of luminance (Y) information at full bandwidth and chrominance (UV) information at half bandwidth.

zoom: To scale a display so that it is magnified or reduced on the screen.

Appendix B: Resources

RECORDING CENTERS

Crawford Communications, 535 Plasamour Dr., Atlanta, GA 30324; Tel: 404-876-7149, 800-831-8027; Contact: James Brooke; Services: commercial post-production, NTSC-CAV and CLV, other IVD services.

Department of the Air Force, Det 8, 1365th AVS Bldg. 1269, Hill AFB, UT 84056; Tel: 801-777-4955; Contact: Bill Harris; Services: secure facility, Dept. of Defense use only, NTSC-CAV and CLV.

EditDroid Los Angeles, Division of LucasArts, 3000 West Olympic Blvd., Suite 1550, Santa Monica, CA 90404; Tel: 213-315-4880 (Contact: Jim Stanton), 415-662-1000 (Contact: Tom Scott, Robin Veasy); Services: RLV, NTSC-CAV.

EditWorks, 1776B Century Blvd., Atlanta, GA 30345; Tel: 404-325-2289; Contact: T.J. Sharitz; Services: commercial post-production, NTSC-CAV.

Kimball Audio-Video, 6221 N. O'Conner, Suite 100, 6 Dallas Communications Complex, Irving, TX 75039; Tel: 214-869-0117; Contact: Roger Lee; Services: RLV, NTSC-CAV and CLV, other IVD services.

Magnetic North, 70 Richmond St. East, Suite 100, Toronto, Ontario, Canada M5C 1N8; Tel: 416-365-7622; Contact: Doug Mielke and Gordon Stoddard; Services: commercial post-production, NTSC-CAV.

Magno Sound & Video, 729 7th Ave., 9th Flr., New York, NY 10019; Tel: 212-302-

2505; Contact: David Friedman and Paul Sterzel; Services: commercial post-production, NTSC-CAV.

Optimus, 161 E. Grand Ave., Chicago, IL 60611; Tel: 312-321-0880; Contact: John MacDonald; Services: commercial post-production, NTSC-CAV.

The Post Group, 6335 Homewood Ave., Hollywood, CA 90028; Tel: 213-462-2300; Contact: Randy Gladden; Services: commercial post-production, NTSC-CAV.

The Post Group at Disney/MGM, Roy O. Disney Production Center, Lake Buena Vista, FL 32830; Tel: 407-560-5600; Contact: Rachael Frimer; Services: commercial post-production, NTSC-CAV.

Laser Edit, 540 N. Hollywood Way, Burbank, CA 91505; Tel: 818-842-1111; Contact: John Torcello; Services: commercial post-production, NTSC-CAV.

Telstar, 29 W. 38th St., New York, NY 10018; Tel: 212-730-1000; Contact: Angela Murphy; Services: commercial post-production, NTSC-CAV.

VIDEODISC REPLICATORS

Disc Manufacturing, Inc., 1120 Cosby Way, Anaheim, CA 92806; Tel: 714-630-6700; Contact: Wan Seegmiller.

PDO (Philips and Dupont Optical), 1409 Foulk Rd., Suite 200, Wilmington, DE 19803; Tel: 800-433-DISC; Contact: Kathy Justison.

Pioneer LaserDisc Corp., 1058 East 230 St., Carson, CA 90745; Tel: 213-492-9935, 800-526-0363; Contact: Linda Johnson.

Sony Corporation of America, Sony Dr., Park Ridge, NJ 07656; Tel: 201-930-1000.

TechniDisc, 2250 Meijer Dr., Troy, MI 48084; Tel: 800-321-9610; Contact: Judy Harnois.

3-M, 3-M Optical Recording Project, 1425 Parkway Dr., Menomonie, WI 54751; Tel: 715-235-5541; Contact Mike Sterner.

Source: Recording Centers—Optical Disc Corp., 12150 Mora Dr., Santa Fe Springs, CA 90670; 213-946-3050.
Replicators—Richard Haukom, Haukom Associates, 2120 Steiner St., San Francisco, CA 94115-2226; 415-922-0214.

Index

ABOUT THE AUTHOR

Martin Perlmutter graduated from Harvard College in 1968. He was involved in founding the Phoenix newspaper in Boston. In the fall of 1970, he set up Ghost Dance. From its beginning, Ghost Dance focused on interactive uses of video. In 1971, Ghost Dance opened the Harvard Information Transfer System with a two-way transmission from the Paik-Abe Video Image Synthesizer at WGBH-TV to a classroom at the school. Ghost Dance went on to pioneer music-image television, and to develop devices for portable teleproduction. Before 1975, the company consulted to both CBS and RCA on advanced video technologies.

In the mid-1970s, Mr. Perlmutter was given a Visiting Specialist grant from the National Endowment for the Arts to design and build an interactive television exhibit at Boston's Museum of Science. The result was *Vision and Television*, a popular and durable feature of the museum from 1977 through 1982. The exhibit modeled processes of visual perception, using TV, and then offered participants the chance to create and control images in a wrap-around video synthesizer environment. Following interactive exhibit work in Boston, Perlmutter helped to design the Mobile Science Center for the Lawrence Hall of Science and later did both exhibit design and award-winning videotape work for the New York Hall of Science.

In 1981, Mr. Perlmutter began to produce laser discs, working with colleagues at Videodisc Publishing, Inc., in New York. In 1981, Ghost Dance and VPI produced the MysteryDiscs. Later, through VPI, he produced discs for AT&T and IBM, among other corporate clients. In the late 1980s, he independently produced several discs in Japan for Pioneer LaserDisc including an award-winning Anthology of American VideoArt. Today, Mr. Perlmutter is producing titles for Commodore CDTV, and publishing video reports on major multimedia conferences. He is producing a series of interactive videodiscs on jazz. Using an approach termed Concept Illustration™, Mr. Perlmutter is creating an interactive mathematics teaching package. He consults; publishes a newsletter (*The Green Sheet*); does water colors; and loves to travel with his wife, Miki, and his daughter, Sasi, with whom he recently spent a year, circling the globe.